Brady

F   )VER 1999

SPEAKING OUR MINDS

LISA SNYDER, LCSW

"Lisa Snyder skillfully weaves a tapestry of what Alzheimer's disease is really about by reporting on conversations with diagnosed people. By reading these, and Snyder's commentary, fellow travelers, caregivers and health care providers will find further insight into Alzheimer's disease."

> —Michael Livni, Member of the Executive Committee of Alzheimer's Disease International

"Lisa Snyder performs an important service in this excellent introduction to the experience of persons with Alzheimer's disease. She allows them to speak for themselves as they describe their challenges and hopes. Her interpretations of their voices are consistently thoughtful and illuminating. The reader, then, is effectively brought into the real world of the lives of the deeply forgetful."

> —Stephen G. Post, Ph.D., author of *The Moral Challenge of Alzheimer Disease*, and Professor of Biomedical Ethics, School of Medicine, CWRU

"This is the best book I've ever read on Alzheimer's disease. . . . Ms. Snyder's work teaches us a deep respect for the uniqueness of each individual with Alzheimer's disease; and that the most profound way to learn is to listen."

> —Robyn Yale, LCSW, Clinical Social Worker and Consultant to the Alzheimer's Association

"A sensitive and compelling view of the perspectives, symptoms, issues, and personal reactions of individuals diagnosed with Alzheimer's disease. . . . This is a book to be read, reread, and passed on to others."

> —*The San Diego Chapter Alzheimer's Association Newsletter*

"An important book for anyone whose life has been touched by Alzheimer's."

> —*Feminist Bookstore News*

"[SPEAKING OUR MINDS] offers a different perspective about Alzheimer's disease. It can give social workers, physicians, and other professionals a small sense of what someone feels who suffers from this disease. . . . More importantly, [it] gives encouragement and permission to discuss the diagnosis, changes, and feelings with the individual who has been diagnosed with Alzheimer's disease."

—*NASW California News*

"One of the men referred to in the book, Bill, used to be a prolific writer and editor who has now lost his ability to write his name and much of his ability to speak. But his wife, Kathleen, reports he's proud of his comments in Snyder's book. 'When people come to the house, he picks up the book and shows it to them,' Kathleen says, 'because if you don't have language, this is an offering of your life.' "

—Denise Nelesen, *San Diego Union-Tribune*

"If you can buy one book for your AD resource library this year . . . let it be this one. The book reminds us all, professionals and families living in the Alzheimer's world, of the importance of listening with our ears, eyes, and heart. It will enrich the life of a person living with AD."

—Inge Gatz, Editor, *Early Alzheimer's—An International Newsletter on Dementia*

"This is a book that is long overdue. . . . The introduction, the interviews, the reflections of Ms. Snyder all weave together to give us this insightful and honest tapestry of seven human beings, and of Ms. Snyder herself."

—*Early Alzheimer's—An International Newsletter on Dementia*

"In addition to offering intimate portraits about the subjective experience of the disease and discussing the medical and psychosocial aspects of the disease, Snyder models how to listen to and talk with people who have the disease. She connects with them in spite of their impaired communication skills and invites readers to do the same. True empathy is the core message here. She knows how to empower people with AD by helping them to define themselves in new ways. . . . She succeeds in fulfilling her stated goal of honoring the varied voices of people with the disease."

—Daniel Kuhn, LCSW, *American Journal of Alzheimer's Disease*

SPEAKING
OUR
MINDS

SPEAKING OUR MINDS

Personal Reflections
from Individuals
with Alzheimer's

LISA SNYDER, LCSW

W. H. FREEMAN AND COMPANY
New York

Cover Design: Terry Parish
Text Design: Diana Blume

Library of Congress Cataloging in Publication Data
Snyder, Lisa.
    Speaking our minds: personal reflections from individuals with
Alzheimer's/Lisa Snyder.
        p.   cm.
    Includes bibliographical references and index.
    ISBN 0-7167-3224-6 (hardcover)
    ISBN 0-7167-4010-9 (paperback)
1. Alzheimer's disease—Popular works.   2. Alzheimer's disease—
Patients—Interviews.   I. Title.
RC523.2.S645 1999   98-47903
362.1'96831—dc21   CIP

© 1999, 2000 by W. H. FREEMAN AND COMPANY

Printed in the United States of America
First paperback printing, 2000

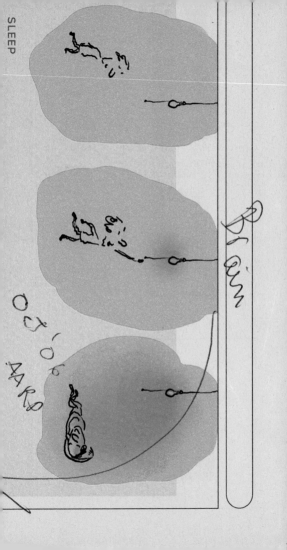

[BULLET

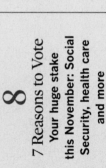

P. 14

8

7 Reasons to Vote
Your huge stake
this November: Social
Security, health care
and more

300 Million a

S ometime in mid-October, the population of the United

Are restless nights making you blue? The answer might be a concentrated dose of light before bedtime—blue light, no less.

For years people have been sitting in front of bright, full-spectrum light boxes to fight off sleep disorders and winter depression. Now scientists have found that doses of blue light work significantly better than white light, and require far less time.

The reason seems to be tied to the many millennia in which our species toiled almost completely outdoors. "We're blue sky sensitive," says Mariana Figueiro, a scientist at the Lighting Research Center at Rensselaer Polytechnic Institute in Troy, N.Y. "We're looking for blue skies to reset our biological clock."

Older people exposed to blue light early in the evening are much more likely to sleep through the night, Figueiro's research shows. The same goes for wander-prone Alzheimer's patients if they get two hours of exposure to blue light. Her findings are consistent with those of Harvard Medical School in Boston and other institutions.

Overexposure to blue light can damage eyes, but a safe and effective dose, Figueiro says, takes as little as 30 minutes, about a quarter the dose of bright white light.

Blue light boxes are already available commercially. They vary widely in quality, however, so get a recommendation from a health professional.

—Reed Karaim

CONTENTS

PREFACE

I wrote this book to illuminate and honor the varied voices of people with Alzheimer's disease and to alleviate the personal isolation that can accompany this condition. Although we have a wealth of valuable scientific, professional, and caregiver literature, literature that addresses the various complex dimensions of the disease, we are only beginning to explore and make public a crucial perspective: the subjective experience of the person diagnosed.

Since 1987, I have worked as a clinical social worker at the University of California, San Diego (UCSD), Alzheimer's Disease Research Center (ADRC). Funded by the National Institute of Aging, this comprehensive research center is one of the original 5 of the now 29 centers across the United States dedicated to the understanding, treatment, and ultimate prevention or cure of Alzheimer's disease. For many years, my work at the research center focused on providing education, counseling, and guidance to people taking care of a loved one with Alzheimer's. We provided the bulk of our services to families, knowing that if we assisted the caregiver, this would ultimately benefit the person diagnosed. Thus, we provided support to the patient through the conduit of the family.

In the early 1990s, with advances in early detection of Alzheimer's disease, patients began entering our research center with only mild impairment. Better able to articulate their thoughts, these participants began to seek information and share their concerns about their condition. It seemed as if a whole new instrument was being introduced to the orchestra of Alzheimer's—one that I was not so familiar with and that warranted more attentive listening. Across the country and, indeed, in other parts of the world, this phenomenon was occurring simultaneously as the voices of people with Alzheimer's disease became more pronounced.

This book evolved out of my investigation into the subjective experience of Alzheimer's disease—an exploration that is still in its infancy. Since 1994, I have been conducting a small series of taped, in-home interviews with people diagnosed with Alzheimer's. I have deliberately chosen people of different ages, ethnicities, and educational and professional backgrounds. All acknowledge their diagnosis and are willing to offer their reflections about the impact of Alzheimer's on their lives. Our conversations cover personal history; diagnosis; the various cognitive, behavioral, and emotional dimensions of the disease; family and social interactions; and philosophical or religious perspectives. I transcribe and edit the interviews to form a narrative. Only in the case of grammatical accuracy or clarity do I alter the wording of each person's unique dialogue.

My professional world is in an academic research setting. This book, however, is not an academic manuscript nor is it based on a study with scientific methods and research outcomes. The emotional, psychological, and intellectual process of conversational inquiry is one whose methods are not readily objectified and whose outcomes are not easily measured. *Speaking Our Minds* expresses the thoughts, feelings, concerns, and experiences of seven persons diagnosed with Alzheimer's disease who hope to teach, lend insight, and evoke understanding in those who are willing to open their minds and listen. I have sacrificed the more definitive conclusions of data for the potential discoveries of dialogue.

In that spirit, many invaluable conversations have nurtured the creation and development of this manuscript. Endeavors involving Alzheimer's disease are rarely solitary efforts. Although the physical illness strikes the patient, family members, friends, and professionals are drawn into that person's world. And so it is with the writing of this book. Although the task has been mine, the ability to see the project through drew on the support, wisdom, and guidance of many people along the way.

My work in the dimensions of Alzheimer's disease would not be possible without the support of UCSD's Alzheimer's Disease Research Center. I am indebted to Leon Thal, M.D., director, and to Robert Katzman, M.D., codirector, for the wisdom, commitment, and indomitable enthusiasm that they bring to their long-standing work in the field. Their continued support of my work at the ADRC and their review of sections of this manuscript have been invaluable.

I extend my ongoing appreciation to the whole multidisciplinary staff of the ADRC. It is a privilege to work with such a caring and skilled group of colleagues. In particular, I must thank Cecily Jenkins, Ph.D., whose astute feedback and sincere support throughout this project helped to shape the structure of the book, and Doris Bower, R.N.C., A.N.P., who, since retiring from the ADRC, volunteers once a week to cofacilitate our patient support group with Cecily and me. David Salmon, Ph.D., was a patient and valuable resource on the complexities of neuropsychology; Douglas Galasko, M.D., and Ron Ellis, M.D., provided helpful consultation on genetics and reviewed sections of the manuscript for accuracy. Phyllis Lessin, retired assistant chief of the ADRC, lent her sincere enthusiasm and administrative guidance at the onset of this project; Sheila Arneson, M.F.C.C., during her brief tenure with the ADRC, helped to develop and facilitate those first patient support groups that opened all our minds.

In UCSD's Alzheimer's Disease Diagnostic and Treatment Center, I am grateful to the late J. Edward Jackson, M.D., for his encouragement to pursue this project. He was a brilliant and compassionate clinician who gave generously of his time and thoughts as this book evolved. Special thanks are due to Kim Butrum, R.N.C.S., G.N.P., for introducing me to Bea and for her unfailing commitment to her patients, and to Susan Shepherd, L.C.S.W., who took me on as a graduate social work intern in 1986 and invited me into the world of Alzheimer's disease.

I am likewise indebted to Robyn Yale, L.C.S.W., for her continuing contributions to a creative partnership in the field of early dementia; Dorian Polson, Ph.D., my mentor through so many layers of learning; and Sue Kirk, M.F.C.C., my consciousness raiser. All have enhanced my vision over the years.

I am sincerely grateful to my editors at W. H. Freeman and Company. Susan Finnemore Brennan saw promise in my initial proposal and participated in an encouraging and constructive partnership with me throughout this project. Ayisha Day provided astute feedback and refreshing candor. Alecia Mar.zullo, editorial assistant, was always helpful with my numerous inquiries. I owe thanks to Mary Louise Byrd, project editor, for her essential organization and direction in the evolution of this project from manuscript to finished book, and to Diana Siemens, copy editor, for her keen eye for the critical details of language. Terry Parish, cover designer, gave generously of his inspiration and patient collaboration.

I received encouragement from friends throughout this project. I offer my deepest gratitude to Steven Ornish and Marty Chapman Ornish, who gave generously of their professional skills, personal insights, and enduring friendship every step of the way. Brian Holmes, Vicki Austin-Smith, Greg Smith, Lee Knight, Scott Smith, Bea Burch, Lizzie Byrd, Lana Wilson, Kent Wilson, Gwen Fieldhouse, Betty Backus, Michael Longuet-Higgins, and Gail White provided helpful feedback and much appreciated moral support.

Every day, my work illuminates the impact of family—the one we are born into and the one we create. There are family members to thank: my father, John Snyder, who read chapters of the manuscript with attentive care and whose genuine interest in people has been fundamental in shaping my own; my mother, Jennifer Wells, whose creativity and fascination with life is an ongoing inspiration; to my siblings, Bob Snyder, Kathy Vaughn, and Becky Wong, who have given me so much love over the years; Jeff Wong, who read my book proposal with a keen eye; and every member of my extended family and my Irwin family whose extraordinarily distinct voices have taught me to listen and distill the unique messages each one has to speak.

I extend my deepest appreciation to my husband, Jeff Irwin, who was always helpful in lifting my occasional fog with his patient, insightful listening and who sustained me throughout this project with his creative and joyful presence, and my late great-grandmother Elizabeth Sellon, who, when memory failed, never forgot how to see beauty.

This book would not have been possible without the generosity, courage, and trust of Bea and Joe; Bill and Kathleen; Jean and her family; Bob and Erika; Booker and Brenda; Betty and Kurt; and Consuelo. They have made invaluable contributions to this text and to my life. And to all those with Alzheimer's and to their families, who over the years have been my teachers and spoken their hearts and minds, I give my continued gratitude and respect. They have all given so much. This book is one small offering in return.

SPEAKING
OUR
MINDS

Part One

LISTENING

"We were all going along quite productively in our lives until we were confronted by memory loss, confusion, nervousness, loneliness, and isolation. It's as if you're reading a book and someone has torn the pages out."

The confrontation is Alzheimer's disease; the speaker, a newly diagnosed man describing the insidious but dramatic interruption in the progression of his life. Alzheimer's disease is disrupting the life stories of as many as 4 million diagnosed individuals nationwide—people whose biographies were proceeding page by page until the onset of disease interrupted the text. Their lives continue on, but what do we know of their stories? Although Alzheimer's disease is increasingly in the limelight, the person diagnosed with it can live in shadow—an autobiography continuing to unfold with no one writing the narrative.

As physicians identify Alzheimer's at much earlier stages in its progressive course, a growing number of people are living with the disease—and, contrary to popular perception, they are not "senile." These individuals are negotiating a world in which they may still be very active participants, even though their roles are changing. Although they look the same outwardly, the inner reflection of their basic self is taking on a different shape, contoured by disruptive shifts in memory, perception, and ability, and yet highlighted with encouraging moments of wisdom, humor, and insight. Without the personal reflections of the people with Alzheimer's themselves, there are missing pages in the text of our understanding—the pages that tell us what it is like to have this disease, to feel it day to day, to cope with its impact.

Definitions

Most people are acquainted with the word *Alzheimer's*. It's difficult to pronounce, even harder to spell, and it sounds ominous. We hear it on television and the radio; we see it in newspapers and magazines. Sadly, we may associate it with a person we know or a tragic figure we imagine, and when our memory falters we joke nervously that we might be next in line.

There are usually many different ways to define a word, and Alzheimer's has many interpretations. Fundamentally, Alzheimer's disease is a progressive, degenerative brain disorder that occurs gradually and results in profound memory loss; changes in behavior, thinking, and reasoning; and a significant decline in overall functioning and ability. These losses result from the death of nerve cells in the brain (neurons) and the breakdown of the connections between them (synapses) that enable the brain to transmit messages. The course of Alzheimer's can span 2 to 20 years, with great variability in symptoms and rate of decline.

Beyond this basic definition, however, Alzheimer's disease carries many different meanings. The impact of the disease is far-reaching, spanning the realms of science, politics, society, family, and—central to this book—the person diagnosed. Each sphere, although united by the common denominator of the disease, defines and experiences its impact in different ways.

In 1904, the German physician Aloise Alzheimer made a name for himself and a disease when he discovered the cause of the progressive dementia that devastated one of his patients in her fifth decade of life. Upon examination of her brain at autopsy, he discovered large numbers of neurofibrillary tangles and neuritic plaques that were responsible for destroying her brain cells. Tangles consist of a mass of twisted, intertwined filaments of the tau protein that are lodged inside the nerve cell and contribute to its disintegration. Plaques are composed of the amyloid protein, which sits outside the cells; amyloid plaques become the depository of decayed nerve endings and other residue from dying cells. These microscopic changes became the hallmarks of Alzheimer's disease; their origins and devastating effects remain a controversial and essential preoccupation of many researchers to this day. Although there are myriad other pathological and neurological dimensions of the disease, to scientists, plaques and tangles are the fundamental defining feature of Alzheimer's.

Sociologists, politicians, and those concerned with public health and policy do not focus on the molecular structures of the brain but on the changing composition of society. They rate Alzheimer's disease as the fourth leading cause of adult death and the third most costly disease (after heart disease and cancer) in the United States. It is a significant health problem and a growing expense.

Our life expectancy has increased considerably since the turn of the century, and it continues to rise; the risk of developing Alzheimer's disease grows dramatically with age, doubling every five years beyond age 65. At present, approximately 13 percent of the people in the United States are 65 or older, but by the year 2025, it is likely that this percentage will increase to 20 percent. Currently, those over the age of 85 form the fastest growing segment of the population over age 65. Research indicates that about 20 percent of the population will have Alzheimer's disease by age 85, and up to 60 percent will have it after age 95. Thus longevity, though a hopeful prospect, is also cause for concern. As the current generation of baby boomers grows older and lives longer, Alzheimer's disease could overwhelm our health care system and bankrupt both Medicare and Medicaid. Those responsible for budgeting and financing the costly medical and long-term care for this rapidly growing segment of our society are worried. For them, these calculations and statistics become an alarming and defining feature of the disease.

And what of our collective society and the meaning we ascribe to Alzheimer's? We are an industrious culture—one that values youth, productivity, ingenuity, capability, and independence. We are, in theory, rational. Given the tumultuous impact of Alzheimer's on these desired attributes, what do we value of the person living with this disease? What place do we give these individuals in our society?

When we define our distinctions from people with Alzheimer's, we avoid having to examine our common ground. Although the majority of people will not develop this disease, it is quite likely that at some point in the course of our lives, we will be faced with a temporary, progressive, or permanent condition that will threaten our autonomy, capacity, and functioning. We can distance ourselves from Alzheimer's disease but not from the risk of dependency, disability, and separation from mainstream society that it represents. It behooves us as a society to examine our approach to the care of people with Alzheimer's because the extent to which our communities embrace those afflicted is the extent to which we can ensure our own individual care when we are temporarily or permanently in need. While we strive to eradicate Alzheimer's disease, we must work to cultivate compassion for our everlasting human vulnerability—a prevailing condition that we all share.

Personal Meanings

When Alzheimer's touches the life of someone we love, our definition of the disease is more personal: it is, in essence, experiential. It is defined by the myriad emotions that evolve and transform over the course of the disease, and it is imbued with images of the one diagnosed—the ways in which we once knew that person and the ways our loved one is becoming known to us now. It is the risk of isolation when it seems no one really understands; it is the solace of supportive fellowship when you find someone who does. It is the logistics of getting through the day; it is the sorrows and satisfactions that can accompany each moment. And on any given day, the personal definition of Alzheimer's disease may undergo as many transformations as the course of the disease itself.

For every person with Alzheimer's, there are countless others in close proximity who feel and live with the impact of the disease. Although this book speaks to the direct experience of the person diagnosed, the concerns of the family and friends are never forgotten.

An Unfolding Interpretation

Perhaps the definitions that have been most private, and therefore most elusive, are the personal meanings of Alzheimer's disease to the individuals diagnosed with it. It is common for us to wonder, "What are they thinking?" But it is less common that we actually ask. And in all fairness, there may be complications when we try.

Unlike most other diseases, Alzheimer's is a disease of the brain; it attacks the very organ that houses the understanding, awareness, language, and expression necessary to comprehend a question, access insight and pertinent corresponding information, formulate an answer, and express a response. For some people, the pattern of disease in their brains may render them less aware of the daily impact of the disease. Others are acutely aware of the extent of the changes. Hence, the response to a diagnosis of Alzheimer's can range from complete denial of a problem to an elaborate explanation of the condition. We cannot always distinguish between a denial based in psychological defense and one rooted in neurological impairment. Sometimes it is a bit of both.

Thus, a personal definition of Alzheimer's disease is as varied as the disease itself—as unique as the particular course it runs in each person who has it. Although the objective neurological findings of plaques and tangles are absolute factors in describing Alzheimer's, they are not what forms personal definitions. Absolutes provide a skeleton, but it is the variability that rounds out a person and creates the unique dimensions that will shape who they become and how they cope. In this realm, there are far fewer certainties.

Over the years of hearing countless families express their experiences, concerns, and strategies for coping with Alzheimer's, I have been struck by the uniqueness of each person's voice. Although there are many common themes, the stories are never told in quite the same way. My work has taught me to listen to each story as if it were new—as if it had never been told before. Indeed, the people telling these stories have never experienced anything like their encounter with Alzheimer's disease.

Diagnosing Alzheimer's Disease

As we work to alleviate or eliminate the problems of Alzheimer's, scientists continue to develop ways to detect and diagnose the disease more accurately. There are more than 70 different causes of dementia; Alzheimer's disease accounts for approximately three-quarters of all cases. But to make a confident diagnosis, a physician must rule out any other possible causes of changes in memory, thinking, and functioning. This comprehensive evaluation is essential: although we cannot yet halt the progression of Alzheimer's disease, many other causes of dementia are reversible or responsive to treatment.

Because Alzheimer's disease can affect the patient's ability to convey the extent of his or her symptoms, physicians often consult a spouse or close family member to provide a more detailed description of the onset of memory or functional problems. This is followed by a physical and neurological exam of the patient. A CAT scan or MRI of the brain is ordered to rule out stroke, tumors, or brain changes attributable to other diseases; mental status questioning or more comprehensive neuropsychological testing enhances diagnostic accuracy by evaluating specific patterns of memory loss and other cognitive impairment; and an assessment of mood is essential to rule out the changes in memory and concentration that can occur with depression.

Aside from these fundamental procedures, some physicians may order a lumbar puncture to rule out certain brain infections or to look for elevated levels of a specific protein associated with Alzheimer's disease. Blood work should be included to rule out kidney, liver and thyroid malfunctions, nutritional deficiencies, or metabolic imbalances. Blood work may also include evaluation of a specific form of the protein ApoE, which is associated with increased risk of Alzheimer's disease.

Although, at present, an absolute diagnosis of Alzheimer's disease can be made only by a brain biopsy that reveals the hallmark plaques and tangles, physicians can achieve 90 percent accuracy in diagnosing Alzheimer's by performing a thorough evaluation that does not include this invasive procedure.

As You Read These Pages

With the exception of Consuelo, whose circumstances will become clear, the individuals profiled in this book have all received comprehensive evaluations to determine the diagnosis of Alzheimer's disease. If you are reading this book and have been diagnosed with Alzheimer's, you will be introduced to seven of your peers. Some of their voices may resonate with you and speak to your thoughts, feelings, and reflections. Other voices may be discordant or offer less with which to identify. Such is the nature of honest discourse about a highly variable disease. Alzheimer's disease will profoundly affect you, but it will not wholly define you. You will have your own unique voice as you feel the presence of the disease in your life. As you listen to these individuals speak their minds, through your own dissension or agreement, may you find a forum for your own voice to be heard.

If you are caring for a person with Alzheimer's disease, this book is not intended to be a manual or practical guide. There is no advice offered, no steps outlined. Such books are invaluable, and your local chapter of the Alzheimer's Association can recommend a number of excellent references. Rather, this book will provide insights into the concerns and experiences of your loved one and will illustrate the profound importance of your own role in his or her life. The issue of human relationships is ever present in these narratives, revealing the ways in which we open our hearts and minds to one another—and how we sometimes shut them down, too. You may want to read this book with your loved one and use it as a starting point for dialogue. If the one in your care denies having the disease, I commend you for seeking the voices of those able to acknowledge it so that you may learn from them what your loved one is unable to convey.

Professionals working in the field of Alzheimer's may or may not have direct contact with those diagnosed with the disease. Work with this disease requires a breadth of disciplines, from the molecular biologist in a research laboratory to the personal care attendant in a nursing home. Regardless of the skills we bring to the far-reaching realms of Alzheimer's, all efforts are ultimately motivated not by the disease itself but by the people living with the disease. If Alzheimer's disease were a benign condition, it surely would not galvanize our attention.

And whether our attention is directed to science; medicine; social, psychological, or spiritual services; ethics; policy; advocacy; or administration, this book reminds us of the people who form the core of our commitment.

Perhaps you are reading this book because you know someone with Alzheimer's or someone who is caring for a person with Alzheimer's. Maybe the disease has not yet personally touched your life, but you are interested in understanding more about it. We are all invested in the function of memory—in cultivating its development and maintaining its capacity. It is central to our daily life, and Alzheimer's disease poses a considerable threat to this cherished possession. Indeed, many people spend a great deal of time and money on memory-enhancing dietary supplements, workshops, or training programs aimed at boosting the brain's memory function. Because of the centrality of memory, Alzheimer's disease is a very frightening prospect. Beyond any personal insight you may gain from this book into the profound impact of the disease on memory and identity, you make an invaluable contribution to the field of Alzheimer's by trying to open your mind to the experiences of others. We do not all have to be directly involved in Alzheimer's to participate in creating a climate in which, ultimately, knowledge, awareness, advocacy, and compassion can grow to override this devastating disease. It is a collective, worldwide effort, and every citizen counts.

These Seven People

Although diagnosed with the same disease, the people profiled in this book experience different neurological and psychological dimensions of Alzheimer's that engender personal definitions of the disease. Their varying ages, ethnicities, backgrounds, religions, and coping styles illustrate the broad spectrum of humanity affected by Alzheimer's. But, as multidimensional as their stories are, these seven people cannot speak for the whole Alzheimer's experience. As they speak with honesty, candor, insight, and uncertainty, we learn that there is no one way to go through this disease—seven other people would tell us seven different stories.

Although differences are distinguishing, the themes that permeate these messages are unifying: the moment-by-moment encounters with memory loss; the ambivalence about disclosing the diagnosis to others; the concern about being a burden to loved ones; the responses of family, friends, and strangers; the struggle between dependency and autonomy; the inability to do the things you once enjoyed; the ability to laugh and find humor in such serious circumstances; the pervasive presence of hope.

As these themes unite each person profiled, they also serve to unite the reader with aspects of the Alzheimer's experience. For these concerns are not necessarily limited to Alzheimer's disease. They are unavoidable concerns we are all likely to encounter at different times throughout life. Thus, this book is also about dialogue—about common ground for a meeting of minds. The reader is invited to listen to the messages of those who speak here—to recognize the differences and to honor the similarities. In this way, the often isolating world of the person with Alzheimer's disease becomes, to the extent possible, our collective world. A society becomes a little more sensitive and a person with Alzheimer's disease a bit more connected.

Although the individuals profiled in this book, to a large extent, are able to articulate their reflections and feelings verbally, in all likelihood their expressions will change over time from verbal to nonverbal—to more behavioral or symbolic gestures. We rely on language as the primary vehicle of communication to bridge minds. And when behavior begins to speak, it can seem like a whole new language—one fraught with confusion and frustration for both the sender and the receiver of the message. The world of the speaker and the world of the listener begin to feel separated by the disease.

If we can learn the themes of communication early on, perhaps we can be sensitive to the ways they might be repeated later in the course of the illness. People may continue to experience similar feelings but express them differently as their confusion increases and their capacity to articulate decreases. We must begin to listen early in the course of Alzheimer's in the hope that doing so will enhance our understanding as the disease progresses. The variability in how people will continue to express their minds is testimony to the uniqueness of each person before this disease strikes and long after it has shown its dramatic effects.

In that spirit, this book is also about identity. And it is about hope. It is about people who are defining themselves and a disease in new ways, not only by what has been lost but also by what is enduring.

Part Two

SPEAKING

Bea

"Sometimes we can laugh about all of this instead of crying, but I wouldn't wish this on anyone."

Bea came into my life well-recommended. She was followed in our UCSD Seniors Only Care (SOCARE) clinic by a nurse practitioner who was aware of my interest in interviewing individuals diagnosed with Alzheimer's disease. After a three-year history of memory problems beginning at age 72, Bea had recently been diagnosed with Alzheimer's. Seeking any treatment available, Bea transferred into the SOCARE clinic to participate in an experimental drug study aimed at slowing the progression of her disease.

Bea's previous medical records were effusive in their assessment of her "most pleasant" disposition. Her gentle, cooperative temperament was every physician's dream, and her unassuming manner could easily be attributed to a somewhat passive disposition. But perhaps it was Bea's benign demeanor that made her ultimate commentary so compelling. Our nurse noticed Bea's capacity for candor and hoped that further discussion would evoke greater appreciation for this pleasant woman's more concealed dimensions.

Bea was raised in rural Arkansas and attended night school to obtain her high school equivalency before securing long-term employment with a nationwide bakery. Twice widowed, she was wise to the demanding challenges of loss, grief, and recovery. The death of her second husband prompted her to seek a new life in San Diego, and in mid-1980, Bea packed her dog, cat, and belongings and moved west. A decent retirement pension allowed her to stop working, and since she had no children of her own, she volunteered at organizations serving infants and brain-damaged children. "I was driving a car and getting around and getting independent," Bea recalled with an amused smile, "and then along came Joe." Joe became Bea's third husband.

After passing the green undulating carpet of a small community golf course, I approached their double-wide mobile home, where Joe met me on the walkway. He was protective of Bea, but his confidence in our nurse practitioner established the trust necessary for me to conduct my interview. His welcome was friendly as he ushered me into their home, where Bea greeted me brightly.

Joe and Bea's spacious living room housed two matching recliner chairs, to which they immediately gravitated. The habitual nature of their movements made me think that they might be a couple who derived security from simple routines and rather well-defined roles. I was curious about how their relationship began, and as Joe sat in witness to the conversation, I asked Bea how they met.

There was a luncheon at the senior center and I brought the lemon tarts. Joe asked how many lemons it took to make a pie. I said, "Bring me the lemons and I'll make you the pie!" So he sent me a bag of lemons. On Saturday I made him a lemon pie, and on Sunday when I went to church, I took it to him. That was the start. We didn't go together very long—about two weeks—and then he said we might as well save all this wear and tear going back and forth and just get married. So we've been married nine years now.

There was a tone to Bea's story that was both sentimental and sensible. Her feelings, whether painful or pleasurable, were easily expressed. Yet she also possessed a practical side that allowed her to respond to these sentiments directly and matter-of-factly. This adaptive dichotomy was present throughout much of our discussion.

First Signs

Bea was mystified as she recalled the startling and peculiar behaviors that marked the onset of Alzheimer's disease more than three years before.

One day I was driving into town to get my hair done. I was supposed to make a left-hand turn but instead, I went straight ahead into oncoming traffic. A policeman pulled me over, and I told him that I didn't know why I did it. I'd driven this same route all along and never had any problems. It was so strange. It was terribly scary. I could have really hurt someone. That was the first time I knew there was something wrong. I haven't driven again since then.

Also, I used to volunteer at church, filling in the greeting slips in the pews. I'd do one row but then I'd go to do it all over again. Then when I went to the ladies' room, I came out a different door and I was lost. It took me a minute to get my bearings in my own church. I didn't have any idea what the problem could be.

The doctors put me through every test imaginable, and I was diagnosed with Alzheimer's disease. It was something that I wouldn't want to go through again. The last person who interviewed me was the neurologist. He was very indifferent and said it was just going to get worse. It wasn't professional as far as I'm concerned. If he had just shown a little compassion. He was there to diagnose my problem, but he wasn't there to understand my feelings. He had no feelings for me whatsoever. I've hated him ever since. Health care professionals need to be compassionate. That's all there is to it. There but by the grace of God go I.

Bea's traffic incident, her repetition of the greeting slips, and her disorientation all revealed the most common initial symptom of Alzheimer's disease: memory loss significant enough to alter one's previous level of functioning. Bea's gradually worsening impairments frightened her, and she willingly sought help. But this is not always the case. Some people, while displaying cognitive changes that are clear to others, do not acknowledge a problem. Memory deficits may render them unable to remember their own impairments, or they may be too frightened or ashamed to admit them. Others are aware that their memory and abilities are changing and courageously address these difficulties.

Yet even when people with Alzheimer's disease are self-aware, the medical community has long assumed that they have little insight into their own impairments. As a result, they are sometimes treated in less than compassionate ways. Bea felt dehumanized by her doctor's evaluation. She had been reduced to a set of symptoms that classified her into a diagnostic category. But Alzheimer's disease affects more than the neurological structure of the brain. It impacts the corresponding social, familial, and psychological foundations of well-being. When we fail to mention the assistance that may be available to help in coping with Alzheimer's, we leave no room for the possibility of hope, resourcefulness, and meaningful living in the face of a disease whose name has come to represent futility. Every physician or health care professional *must, at the very least, include a referral to the local chapter of the <u>Alzheimer's Association</u> when giving a diagnosis to a <u>patient and family.</u> The simple yet invaluable phrase that <u>"you are not alone,"</u> although seemingly trite, is all too often absent from the evaluation process.*

Getting Through the Day

As our conversation moved beyond the trauma of diagnosis, it became evident that the demands of daily living were now Bea's primary concern.

One of the worst things that I have to do is put on my pants in the morning. This morning I kept thinking there is something wrong because my pants just didn't feel right. I had put them on wrong. I sometimes will have to put them on and take them off half a dozen times or more. I think I'm putting them on opposite to the way they're supposed to go on. It's so frustrating because I look at them and try to figure it out. I think I know the way to do it and I put them on and it's wrong again. Maybe I'm going to have to get another type of pants.

I keep up with the housework, but I'm not able to sew or iron like I could before. Anyone who can iron a white shirt can't have Alzheimer's disease! Setting the washing machine is getting to be a problem, too. Sometimes I'll spend an hour trying to figure out how to set it. I can start out with a bang, but by the last load I'm so confused. If I stay with it, I'm fine. But if I go off and do something else and come back, I'm lost.

It seems like Joe's having to do too much, but I can't do anything about it. He has to do all of the cooking. He fixes all the meals. I make the salad and set the table and do what I can to help. I used to do a lot of entertaining, and I'd take pride in setting a nice table. But now I don't even know on which side the fork or knife is supposed to go. I'll get the plates on and then I'll get the silverware. I'll ask Joe, "Where do these go?" He'll show me, but then I don't remember the next time. That's frustrating. Sometimes he must think I'm awfully stupid. I feel so dumb when I ask, "Where is my fork?" He'll answer, "It's right there." I've put it on the table half the time. It's just weird. The simplest things that I've done before, I can no longer do. Sometimes I can do them, though, and then I think he wonders if I'm just putting all of this on.

There were times when Bea's life was like a game of charades. She had clues—a gesture here, a word there—but she expended considerable energy trying to grasp the whole phrase. She turned the clues over and over in her mind, rearranged them, and struggled to complete a given task. Sometimes the pieces all fell into place and were illuminated by a spell of clarity, so that she could act and succeed triumphantly. But the next phrase would present a new set of challenges. The game didn't become any easier with practice.

This fluctuation in ability is confusing to both patients and families. The inconsistencies may be inaccurately attributed to laziness, lack of concentration, unwillingness, or—as Bea feared—a farce. Yet this day-to-day variability is not uncommon in Alzheimer's, and it contributes to the feelings of disbelief so pervasive in the earlier stages of the disease.

Seeing Is Not Always Believing

Bea also struggled with one of the more perplexing symptoms of Alzheimer's disease: visual agnosia. Although her visual acuity was good, she was losing the ability to recognize or identify what she saw. Hence, she could look directly at her knife, see it clearly, but not be able to decipher its meaning. This agnosia was combined with the more common problems of impaired depth and spatial perception. As Alzheimer's dis-

ease advances in the brain, it can affect the ability to judge depth and distance effectively. It may be hard to tell the height of a step or the space between oneself and another person. One may reach for an object only to find one's hand landing inches away from it. These visual and perceptual disabilities combined with memory loss to create episodes of almost theatrical chaos for Bea and Joe.

On Saturday night before we go to bed, it's hilarious. We get our showers and then I get my clothes laid out for church tomorrow so at least I can know what I'm going to wear. I'll go through my jewelry box and think, "Well now, I have everything just right." But then I won't be able to find my pearls. So, we'll have to start looking. Joe, bless his heart, looks and looks. But it's a real chore, believe me. I make an effort to put everything where I think I know it is and then I go back and it's not there. I probably moved it. It's frustrating because I get so aggravated before I do find what I'm looking for and I think, "Oh just forget about it!" It makes you think you're crazy. You think anybody with any sense could find some pearls. Sometimes what I'm looking for will be lying right in front of me and I won't see it. I don't always misplace things; they're right there, but I just don't recognize them. That's a real problem.

And money! Money is getting to be terrible. I just don't want to handle money anymore because I can't identify it. I can trust my beautician not to cheat me. But when we go out anywhere else, Joe has to do all the buying because I just don't have the ability to figure it out anymore.

It's not difficult to identify with parts of the Alzheimer's experience. It can take an inordinate amount of energy for anyone to find a missing object. The difference lies in the array of strategies available. Normally, we can retrace our steps and try to remember the last time we saw the missing item. For the person with Alzheimer's, this memory of the recent past is likely to have been erased by the nature of the disease. We usually look in the places that are logically associated with the article. But Alzheimer's disease blurs associations to the extent that a bread box and a jewelry box can seem one and the same. Pearls may end up in the pantry. Many people have had the experience of scanning right over a missing object as a result of distraction or harried frustration. But Bea, even when presented with the object, could not always identify it. She could describe the small white beads strung together, but were they her pearls? She might be able to recognize the coins in her pocket as money, but which one was the quarter and which was the dime? Any mathematical calculations needed for money transactions were nearly impossible.

It can be terribly alienating to recognize that what is common understanding to most people has become an enigma to one's own self. Bea felt like a foreigner in her previously familiar culture, and she sensed that the members of her community were more aloof than welcoming.

They Might Catch It

We've never tried to hide that I have Alzheimer's. But everyone acts like they don't want to get near because they might catch it. They don't know how to deal with it. I was always so social before, but now I don't like to be around people. People don't want to talk about it. I don't have a friend anywhere who I'm close to that I can communicate with. I don't even like to talk on the telephone anymore. I take a message and I'm afraid I'll forget part of it before I have written it down. So it's a lonesome old world at times.

A woman at church had a husband with Alzheimer's and he would run away. She would be frantic to find him. The other day, someone asked Joe if I had run away because that's what her husband used to do. But I don't have any desire to go.

I used to go around the block but I don't like to do that anymore because I'm afraid I might get confused. Two or three times a day I get up and walk around the house or go outside and pick up the yard a bit. But I'm a coward. I don't want to get away and have something happen.

Bea often felt ignored. Alzheimer's disease acted as a veil that separated her from others, and she felt nearly invisible in social settings. Indeed, Bea's memory loss created challenges in her relationships, and as she perceived others' discomfort with her disability, she also retreated from their company. But sometimes the veil is over the eyes of the beholder. Our own biases, past experiences, or naiveté about the disease shroud us in avoidance or discomfort; our self-imposed barrier effectively screens out opportunities for continued relationship with a person who has Alzheimer's.

At church, people offered to shake Bea's hand, but because of her visual-spatial problems, she was unable to respond: she could not locate the greeter's hand accurately in space. As a result, the common misperception was that Bea had become too impaired to recognize the appropriate response to a handshake. The truth, however, lay in a painful predicament: Bea knew the rules but was unable to comply with them. A cycle of responses ensued where assumptions were made about the severity of her overall impairment, and she was then treated as if the assumptions were true. This only increased her feelings of separateness. As a result, she became more emotionally and physically dependent on Joe and placed him at risk of being isolated, too.

Poor Joe, he's stuck with me all of the time. I try to get him to go down in the afternoon and play pool and get away for a few hours because otherwise he's with me constantly. We used to go to a Sunday class, but I don't participate anymore. I have no concentration whatsoever. I don't feel comfortable, so I don't go. I was very active before. We used to put on dinners, and if someone wanted a volunteer, I was always the first one to offer. But I'm not dependable to do anything like that now. I depend on Joe for everything. I'm isolating him as well as myself, and I'm not being fair to him.

I'm satisfied to be here at home. I'm not really bored with my life. I've always had the feeling that you could be as happy as you want to be or as miserable as you want to be. I'm fortunate to have the man I have to go through this with me. I couldn't manage without him. It wasn't easy at first to ask for help, but now I ask for a lot of it. Sometimes I think Joe should get so tired of this, but he has more patience than I have. Sometimes we can laugh about all of this instead of crying, but I wouldn't wish this on anyone.

Not long into our conversation, Bea began to weep. I chastised myself for eliciting her tears, but Bea was nonchalant. "I've always been an easy crier," she smiled with a familiar and comfortable resignation. Indeed, as we spoke, her tears of sadness and her tears of laughter all flowed into a very watery experience that seemed quite familiar to her. Joe went to find more Kleenex with a matter-of-fact recognition that this might be a two-box meeting. There was something profound in all this, and I sat quietly treading water amid her fluid emotions. As we approached each loss in Bea's life, swells of sadness would break into waves of laughter. Bea possessed an extraordinary ability to be entertained by the absurd. When she thought about the overarching degenerative implications of her disease, she was very depressed. At times, she said she just wished she'd die. Then, inspired by a moment of satisfaction or humor, she expressed contentment with life. It was striking that she experienced her condition as neither tragic nor comedic; it was both—and sometimes, simultaneously.

Bea sensitively recognized that the burdens of her illness could have a significant impact on Joe's well-being. The role of the care provider is fundamental in the life of a person with Alzheimer's disease. It is estimated that 7 out of 10 people with Alzheimer's disease reside at home, with almost 75 percent of the care provided by families or friends. The health and well-being of these caregivers is critical to the health and well-being of the person with Alzheimer's. Many care providers also report feelings of abandonment or separation from their previous social and support circles. Some experience increased depression, insomnia, or fatigue as they try to come to terms with their loved one's condition. There are organizations that provide outreach to families and serve as essential resources for any care provider (see appendix). Although old sources of support may retreat, new ones can advance to ease the burden in this challenging time.

A Stronger Faith

Although the church was a disappointment in addressing Bea's emotional needs, it did provide a spiritual structure that she considered fundamental to her ability to cope.

I don't know what I would do if I didn't have my faith. That's the only thing that holds me together. I know that all I have to do is call on the Lord and he's right there. I ask him if it be his will, to take this Alzheimer's away but if it isn't, help me accept it. That's about the only thing I can do. He has a reason for everything although I can't imagine why I got Alzheimer's. I don't know that there is ever any indication of who will get this and who won't. Everything was hunky-dory one day, and then all of a sudden the bottom fell out.

I think I have a stronger faith than I had before. Oftentimes I try to be so human and think that I can do it all myself, and I know that I can't. I just have to live my life to the best of my ability and put it in the Lord's hands.

I don't like to make mistakes. I get aggravated with myself, and I can cry a lot easier than I can laugh. I would rather live with better feelings than I have at times. I'd like to be perfect. But I never have been and never will be. This disease makes you feel so helpless. I've lost my feeling of self-satisfaction that I was capable of doing things, and I resent it a little bit. But I'll live through it.

I'd like to be a good example for somebody coming along. I have accepted the Alzheimer's, but I have to gripe some! I accept it because I can't do anything about it. It's just one of those things that I have to take in stride. I think I have a pretty good attitude. I don't spend a lot of unpleasant hours or days thinking about it. I go on with my life and do what there is to do. And my life is pretty good, all things considered.

After our initial interview, I lost touch with Bea, as she did not participate in our research center. I did receive periodic updates on her condition from the SOCARE nurse practitioner, however. Two years after our meeting, Joe had a significant stroke and was hospitalized for four months. He recovered remarkably well, but he did suffer mild impairment. Bea's only relative, her nephew, arranged for her to receive in-home help during Joe's hospitalization. But with Joe now disabled, the couple would require more comprehensive and ongoing assistance. Upon Joe's discharge from the hospital, they moved into a retirement community that offered assistance with meals and personal care as needed. Dissatisfied with the atmosphere and quality of care, Bea and Joe made a series of moves over the next two years before finally finding an adequate community. Four years passed before I met with them again.

When I saw Bea, I was struck by her frail appearance. She sat woodenly in her recliner. Her faint voice greeted me in a high-pitched monotone, and I wondered if she had suffered a small stroke. Joe sat opposite at a cluttered desk where the bulk of their life affairs were gathered in seeming disarray. I now occupied the recliner next to Bea. The living room of their one-bedroom apartment was dim. The glow of bright sunlight softly illuminated their heavy, pulled drapes, and I resisted the temptation to fling them open. I felt claustrophobic; my initial dismay at Bea's appearance and living quarters began to smother my hopes for communication. I risked becoming yet one more person who exited Bea and Joe's ever-shrinking circle. Reining in my judgments and refocusing on Bea, I effectively redirected my preoccupation with the apartment drapery, reached within, and tugged back my own internal curtains.

I reframed my impression of their relatively sparse environment: Joe and Bea had simplified their lives. Their small apartment was now furnished with only the essentials. I recalled out loud the effort they undertook to find things in their large mobile home. And as we began to review the changes over the last four years, it became evident that although physically frail, Bea maintained remarkably robust communication abilities. She became increasingly alert, and her wispy but resolute chuckle wafted through the previously stagnant space: "We had garage sales and we gave a lot away. Now I don't have anything left to lose!" Amused by her candor, I asked for an update on her dressing problems, and Bea appeared pleased to have procured a solution for that more complex dimension of her daily life.

I still have <u>trouble getting dressed.</u> So, I don't do it myself anymore. I have a girl who comes in and does it for me. It's a relief. She comes in every morning and gets the shower going and takes care of the rest of it, too. I feel so helpless but I can't do anything about it. We knew that I needed to have some help. This girl is precious. She's real sweet. She dresses me and if she's not here, Joe can do it but it's a hassle. I don't know what to do and he doesn't know what to do, so neither of us does anything. It's awfully hard for a man to know what to do. Maybe I should just join a nudist colony!

Gusts of laughter now circulated through the apartment. Reveling in this vitality, I felt extraordinary gratitude to Bea for unwittingly jolting me into awareness of the life force that persistently sought pathways through the disorienting maze of her mind. Bea's functional disability was characteristic of advanced Alzheimer's disease. She needed help with bathing, grooming, toileting, and eating. Physical weakness necessitated use of a wheelchair. In most cases, comprehension and expression of speech would also be significantly affected. Yet Bea retained these abilities, and her intact humor helped to diffuse the recurrent despair over her other losses.

I never in my life dreamed I'd have any trouble getting dressed because I was so used to it. Now, sometimes I put my nightgown on backwards but I don't let that bother me. I can sleep with it on the right side or the wrong side. Just let me get to bed! I like my sleep.

We have another lady who comes in on the weekend. She's getting used to us, but at first she must have thought it all was a mell of a hess! *(Bea grinned as she intentionally converted "hell of a mess" to avoid swearing.)* She defends me if anyone tries to give me a bad time. Sometimes in the dining room there will be some old biddy who thinks everything belongs to her. But my girl will stick up for me. She takes me to the dining room in the wheelchair. I can walk, but sometimes I have trouble with it. My legs just give out. I think the wheelchair is wonderful. When it comes to getting somewhere, you get there a lot faster.

I can feed myself to a point. But I have a hard time identifying the food. I can see it, but I can't tell what it is. So, it's a hassle. Today Joe was helping me with the chocolate ice cream, and he spilled it on me.

Bea

I had noticed the chocolate stains on Bea's white pants. My first response was sadness that she was becoming unkempt and not receiving the caring attention she deserved. But as I heard the tale of the chocolate stains told through the faint but enlivening sounds of Bea's laughter, I began to appreciate that she and Joe managed to find amusement in each other's disability. There was no element of ridicule; rather, the humor was born of empathy. Bea's visual agnosia now extended to her food. She could not identify it well enough to know how to eat it. Joe's stroke rendered him less coordinated. Although he knew the steps to help feed her, he sometimes slipped in the execution.

Although dismayed by her condition, Bea emphasized an appreciation of the help offered to her instead of the personal losses underlying her dependency. She enjoyed the women who attended to her care. Their presence was a bright spot in her life rather than an intrusive reminder of her disability. It is commonly thought that people in the more advanced stages of Alzheimer's disease, like Bea, may forget the nature of their own illness due to severe memory loss. Thus, what seems to be acceptance could be born of diminished awareness. I was curious about Bea's evaluation of her impairments and gently posed a very direct question: "How's your memory these days, Bea?"

"What memory?" (Bea labored to laugh and speak at the same time.) "Sometimes it kicks in, but other times it kicks out! I don't have any difficulty talking though, knock on wood." (Bea knocked on her head.) "But sometimes it seems like everything frustrates me and I want to throw something across the room. So I do. Then I feel like a goof when I have to pick it up."

Bea's mood shifted and she became teary. Joe looked a bit puzzled, so I inquired further: "What do you throw, Bea?" (Bea pointed silently to her wet Kleenex.) "I just don't take it very well sometimes. All in all, I have a pretty good disposition. I don't get nasty often. But when I do, I get awfully nasty."

Although the conversation took place primarily between Bea and me, Joe's presence was attentive, and occasionally he felt compelled to contribute to the discussion.

"I don't see it. She's not hard to live with," he commented, as if to reassure Bea that she wasn't as ornery as she thought she was.

"You're not answering the questions, Joe!" Bea's tone was defiant.

Although their camaraderie was playful, there was an undertone of irritation from Bea. It was unclear how much frustration Bea vented outwardly, but the experience of absolute exasperation is very real for many people with Alzheimer's disease. While Bea's sense of humor and ability to accept help alleviated some of the stress, it is likely that her inner world raged more often than her outer appearance revealed.

Despite their aggravating disabilities, Bea and Joe found pleasure in some of the activities offered by the retirement community, particularly the musical entertainment. Although they could no longer make the trip to church, they did watch Sunday morning services on television, and Bea continued to receive considerable comfort from her faith. Her nephew and his wife lived up the coast many hours away, but they maintained supportive phone contact and made occasional visits. Time passed, and although Bea said that she was relatively content, she remained ambivalent about her life.

I've hoped for a long time now that God would take me. It would be better for me because I would be freed from this mixed-up feeling I always have. But for now, I can't really do anything about it. You have to take life the way it is.

I don't think about the Alzheimer's that much. I guess it's a good thing that I don't. You've got to get it behind you. You just have to think about something else. You keep thinking something is going to happen and it will get better. It doesn't, but you keep hoping. It's a difficult situation, but you have to accept it. Live with it, and try to be happy. That's all you can do. You have to take the good with the bad.

"I think she's done a pretty good job of it." Joe's praise of Bea was genuine.

"Even though you'd like to kick me in the pants once in a while?"

"That too," Joe acknowledged with a hearty laugh.

"You don't get away scot-free, Joe," Bea jested knowingly.

As I left Bea and Joe's apartment, I reflected on the significance of Bea's retort. Truly, Alzheimer's exacts a toll on both patient and family as they encounter the disease's many challenges. Although people often question who suffers more in this dynamic, it is a misguided inquiry. Bea lived with a condition that persistently scrambled any sense of internal structure or stability. She sought to rely on a basic foundation of acquired knowledge: the ability to identify and define the visual world around her; the skills to succeed in fundamental tasks of daily living; the means to access archives of memories. But her disease gradually and persistently wreaked havoc on this foundation, leaving her bewildered and disoriented within the domain of her own mind. And Joe, witness to the impact of Alzheimer's disease on Bea, was continually challenged not only to understand her inner world but to interpret the outer one for her. He was forced to maintain a structure of life for both of them when Bea could not uphold one herself. Bea and Joe each experienced unique and equally significant challenges that deserve acknowledgment and attention.

As I walked toward the exit of the retirement facility, a large lobby established a transition between Bea and Joe's apartment and the world beyond. I moved outside into the bright daylight, spotted my car, opened the door, and sat quietly. I had met with

Bea for about an hour and a half. It was a rich and warm encounter, sad and humorous. It was illuminating. And it is fair to say that it tired us both. We each put forth effort: she through concentrating her increasingly elusive cognitive and physical capacities into an extraordinary ability to communicate verbally about her world, and I through establishing a still space within myself where my initial judgments, fears, and sorrows were quieted to hear her messages.

And it was not just Bea's verbal messages that were important to hear. I am reminded of how quickly our eyes render judgments. Had my evaluation of Bea stopped with my first image of her frail physical presence in the dim room, I would have unknowingly rendered her incapable of the dialogue that subsequently transpired. Four years was a long time between visits, and of course Bea was different. Her visual agnosia was even more severe, and it combined with her progressive memory impairment to make the most fundamental tasks of personal care practically impossible. Whether as a result of small strokes, the progression of Alzheimer's, or both, her body had weakened considerably, necessitating much more assistance with movement. But in spite of these differences, Bea was also very much the same in some ways. Her humor, language abilities, and insight were remarkably resilient.

My interview with Bea pointedly reminded me of how quickly we measure disability, deficits, and differences at the risk of overlooking ability, strengths, and commonalty. Indeed, Bea had experienced this phenomenon in both social and medical interactions ever since the onset of her disease. Despite her verbal abilities, she became more defined by her impairments and less validated for her capacities. Surely, the challenges for individuals with the disease and for those with whom they interact are significant as Alzheimer's progresses. Because of the particular pattern of disease, many individuals may not be able to maintain Bea's insight and verbal abilities. Yet throughout the course of Alzheimer's, each person continues to convey mes-sages through action, gesture, expression, and behavior. The dis-ease does not result in a complete inability to communicate. But it can require our time, energy, receptivity, and ingenuity to observe, listen, and comprehend effectively.

Although Bea felt both sadness and aggravation over her losses, she did not do battle with her illness. Her response could be interpreted as a despondent and passive submission, but this would not credit a deeper purpose for her coping style. Early in the course of her disease, Bea's faith in a divine plan resulted in her decision to try to accept her fate. If she had to have this disease, she hoped to have it gracefully and be an encouraging role model for someone else faced with the same diagnosis. Although the neurological course of Alzheimer's dis-ease can considerably alter the best laid plans, Bea has been spared marked changes to her temperament. Despite consider-able challenge and upheaval, her "most pleasant disposition" has remained intact.

Bill

"As difficult as it is for me to read and write any more or to talk about myself, I think it is important therapy for me and a help to others to know what I am experiencing."

I can't remember the first time I met Bill. Our relationship is like a language I've learned over time with no recollection of my first words. In 1992, Bill arrived, accompanied by his wife, Kathleen, for his first-year evaluation at our research center. At age 54, with a diagnosis of Alzheimer's disease and the progressive worsening of his memory and language and writing abilities, Bill was forced to retire from work he greatly enjoyed as a magazine editor for the Foreign Service Division of the United States Information Agency. The agency represents the government's policies overseas and relies on its magazines to publish articles that illustrate beneficial bridges between the United States and other countries worldwide. Bill recalled his profession with gratitude: "I have an inkling of what Thomas Edison meant when he said, 'I have never worked a day in my life—it has all been for fun.' Nothing compares to that rush of creating something for others, be it inventing a light bulb or publishing a magazine."

In a cruel and ironic twist of fate, expressive aphasia—the inability to produce verbal language—combined with the loss of ability to spell, was among the earlier and more prominent symptoms of Bill's Alzheimer's disease. Language had long been his artistry and expertise, but the composition of words into verbal expression or text was now an arduous and frustrating process. As we became acquainted, however, it was evident that Bill's mind maintained an ongoing narrative of thoughts and feelings about Alzheimer's disease. His friendly, bearded countenance disguised a quality of willful tenacity: he was determined to meet the challenges of his illness head-on.

Despite his language impairment, Bill was very receptive to being interviewed. On the morning of our meeting, I parked in front of his home opposite the community tennis court. Bill and his friend Chuck (also diagnosed with Alzheimer's disease in midlife), were avid tennis players, and Bill was patiently tutoring Chuck in the game. The relationship served as a valuable friendship for both of them, as well as a satisfying source of physical exercise and activity.

I entered the residence through a shady courtyard of greenery and found both Bill and Kathleen waiting to welcome me. I knew Kathleen from previous meetings at our research center and was warmed by her greeting. The interior of their home reflected a deeply peaceful environment. Hindu and Buddhist bronze statuary acquired from their five years with the Foreign Service in India infused the living room with a contemplative and soothing quality. Although life was dispensing considerable doses of stress to the residents of this home, it seemed that if one simply kept inhaling and exhaling in the midst of their living room, the tensions would dissipate and somehow pass on.

Bill had embraced the opportunity to express his thoughts, but now he seemed nervous about the interview process. The tape recorder lay on the table before us, a disconcerting presence. Bill had relied on recorders extensively to compensate *for his failing memory during his last years as a magazine editor. Although the recorder had once been a helpful tool, it now would capture not only the content of our conversation but every struggle and word-finding challenge Bill faced in his current condition. Interviews had been routine in Bill's career, but now he worried about his ability to produce words well enough to articulate his thoughts.*

Bill's manuscript also lay on the table. A year after receiving the diagnosis, Bill began the laborious task of chronicling his life with Alzheimer's disease. He had been obsessed with the project for the past two years and raced to record his efforts to conquer the disease while it simultaneously eradicated his ability to write. Although there was nothing yet recorded on the tape, the manuscript sat with a solid defiance. The substance of black and white print on paper scripted into permanence the ever-increasing elusiveness of Bill's language and memories. Written while his language ability was much less impaired, the text would always stand as a testament to a past identity—an identity transformed by the influence of Alzheimer's. As the tape began to roll, Bill and I used sections from his manuscript as a foundation for our dialogue.

The Onset

Seven years before his 1991 diagnosis, Bill recognized the first episode of memory loss. But because of the family's relatively nomadic existence and the constant stimulus of foreign service work, at first the memory problems were irregular and difficult to comprehend. Bill documented these beginnings in his manuscript.

For 13 of my 27 years with the United States Information Agency, my wife, Kathleen, our two boys, Collin and Neil, several cats, ferrets, and I lived as a vagabond Foreign Service family.

We were posted for two years in Mexico, five years in Austria, another five years in India, and one year in Tunisia.

I think my first major problem with memory loss occurred in about 1984. We had good friends visiting us in Vienna and I became completely tongue-tied. It was so sudden and embarrassing. I was hardly able to show them around because of lapses in my memory. The problem continued for most of their stay. Then it went underground and didn't surface again until several years later in India. I was a duty officer and I had to write the weekly memo that went out to the Washington staff and the three regional branch posts in India. I started to slip a tape recorder into staff meetings because I couldn't take notes or remember. This was also about the time that my spelling went wonky, so I bought a pocket speller. I still need it and use it even for basic words. The problems were very unsettling and led me to retreat into a shell of internal anguish.

When we moved to Tunis, the year before my diagnosis, I tried to learn French. It was at this time that Kathleen became even more concerned for me. I couldn't remember the conjugation of even the simplest verbs. This was just the last in a series of events. I was forgetting my lunch, my briefcase, my keys, and I had to write down word for word what I wanted to say in a telephone conversation. I couldn't get organized and it didn't get any better.

Bill's description testified to the fear and confusion that so often infuse the early stages of Alzheimer's. Many of us have periods of disarray in our lives that do not indicate the onset of disease. Often these periods are attributed to changes in routine, an increase in life stressors, depression, or the demands of new challenges for which we feel unprepared. Once these stressful conditions subside, our organizational abilities return. In essence, the floor is solid underneath the pile of clutter, debris, and spills that accumulate over a period of time, but we must work to clean house.

Yet for Bill, the foundation of his mind was giving way in unpredictable places. As much as he tried to organize his environment, when he walked on previously reliable terrain, his footing was now uncertain and forced him to create elaborate ways to find support and bypass confusing danger zones. His attempts to reestablish order in his life were of limited avail, and the problem became progressively worse. This progression in the face of every attempt to compensate was alarming.

The Diagnosis

Bill's symptoms of memory loss and disorganization could not be localized to any specific cause and began to encompass larger areas. He documents that while he was on rest-and-recuperation from Tunisia, Kathleen and her sister convinced him to be evaluated in a San Diego clinic. After a complete physical and neurological work-up, Bill was diagnosed with Alzheimer's disease. He wrote:

At the time, I was angry with both women and the system in general. I felt like I'd been turned in against my will. We'd been planning to retire after three or four more years in Tunisia, after our youngest son Neil graduated from the University of California in San Diego. But after the diagnosis, retirement was immediate.

At age 54, it seemed like I was labeled incompetent after a lifetime of proficiency. The psychologist who tested me said that I would find it increasingly arduous to work, or even drive a car. I was devastated. After the diagnosis, I remember walking out of the clinic and into a fresh San Diego night feeling like a very hopeless and broken man. The next morning, the neurologist apologized for the psychologist's lack of tact and then went on to critique the salient points of his own findings, reinforced by the bleak pictures from the Magnetic Resonance Imaging

(MRI) showing obvious spaces in my shrinking brain. The scariest time in this whole process was that first diagnosis. I wondered if there was anything for me to live for. It was an awful time.

Many individuals and their families report a similar catastrophic despair when they receive the news. Life feels very frightening, and the future is shrouded in futility. A diagnosis, particularly when given insensitively, can feel very abrupt. The actual progression of Alzheimer's disease, however, is a more gradual process. Bill's impairments were significant and painful obstacles to his work performance and quality of life. He and Kathleen made numerous adjustments in the years preceding his diagnosis, and there would be many more challenges to adapt to as the years advanced. But Bill was by no means completely incapacitated. Alzheimer's disease has many hues. Unfortunately, the diagnosis often extinguishes the light with which to view them.

Alternative Remedies

Bill and Kathleen described a period of darkness that descended and the consequent need to brighten their reality with hope. Seeking to challenge futility, they stopped at a health food store, where they were inspired by a man who had written a book about his success in reversing his Alzheimer's disease. Bill and Kathleen purchased the book and spent the next two years repeating the steps that the author claimed had cured his disease. It was a period of hopefulness during which Bill was convinced he could conquer Alzheimer's.

I tried to purge my body of any toxin that could possibly be causing the disease. I had all of my mercury fillings removed and replaced with granite, and I had chelation treatments to flush any heavy metal deposits out of my body. I started a rigorous diet which included numerous vitamin and food supplements and a lengthy water-only fast. I knew something was wrong, but that just made me try to keep everything right. I thought I could beat it.

I may not be out of the Alzheimer's now, but I understand a lot more about nutrition through my experiences and I think that's important. I am giving my body every opportunity to be healthy and hopefully slow the progression of the disease. I feel that all of the things I have tried have helped in some way and also given me hope. What is most important now is that I am alive, healthy, and aware, and surrounded by my loving family. But I think people are going to recognize that there is more to health and disease than they see now.

Bill's testimony elicited both encouraging and cautious responses from me. I appreciated and respected his drive to occupy his mind with ambitions of health rather than a depressing preoccupation with disease. It was a coping strategy that allowed him to take command in the face of a very unempowering diagnosis. And although "alternative" medicine was once disregarded by Western scientists, there is a growing influx of respect for, and research into, the medicinal value of vitamins, herbs, good nutrition, stress reduction, and physical exercise. Indeed, some of the nutritional supplements that Bill consumed are now being evaluated in research settings as potential treatments for Alzheimer's disease.

But it is imperative that we be cautious and informed consumers. It is often difficult to distinguish between touted remedies based on relevant science, investigation, or effective practice and those that are the products of seductive and disingenuous marketing schemes that prey on our vulnerabilities and our hopes for a cure. When faced with a serious illness and no other hope of relief, it is frustrating to endure the tedious years of investigation and government approval necessary before medicines or treatments are made available to the public. Without formal investigation, however, it is very difficult to determine which remedies are truly helpful, which have harmful side effects, and which are merely a well-advertised drain on financial resources.

Confronted with Changes

Although Bill felt that his experience with alternative remedies enhanced his overall health, the progression of his impairments forced him to recognize the impact of the illness on his daily life. We put aside his manuscript and talked of the changes he encountered now.

Everything I have done up to now in my life has been great. But now I've been confronted with this disease that I can't do anything about. I was an editor and I can't spell anymore. I make mistakes all the time. I can't remember what I want to write or how I'm going to do it. I have to go back and rewrite again and again. I won't think the writing is good, so I'll have to start all over. I can hardly type anymore. I used to do all of my typing throughout my career. Now I'm just typing at a speed of one-two, one-two, and by then, I've forgotten what it is I wanted to say. This disease just comes around and beats me.

I can barely even use the computer. The computer used to be a nice friend of mine, but I can't remember from one time to the other how to set it up. I'm always going to one of my sons so they can set it straight. But I'm wondering what I'll do when they aren't here living with us.

My sons are dealing with this very well. They're good kids. They are always joshing me, and it's fun. They used to learn from me and now I have to learn from them. I don't really like it. I feel like it's topsy-turvy, but that's the way it has to be. I would like to write without having it be such drudgery. I feel so stupid when I have to ask for help. But they don't mind. It's a shame that we have all of this time now and I can't make good use of it. I was so "with it" before. To me, I could do anything. I even built a house. I can't do anything like that anymore.

I know what is happening. I think my disease is getting worse rather than better. I don't want it to happen, but that's the way things are, so I don't pity myself very much—just at the really bad times when I want to kick the cat!

Bill was a versatile man. He possessed the flexibility to live in different countries, adopt various customs, and learn new languages to communicate in his surroundings. But exasperation resonated through his halting speech as he described the dumbfounding circumstances of Alzheimer's disease. The very literary and language skills that Bill had used to adjust to new challenges were now severely compromised.

The worst of my affliction is not being able to speak. When I try to speak, I see what I'm going to say in my mind, and then the words turn around and go further and further out of sight, and I cannot pull them back. Then when I go to speak, they aren't there! So, I don't say anything. It's really a tough grind. I am able to bring thoughts in but I can't get them out. I am forever starting a sentence and finding that I don't know how to end it. And then I hope that the person I'm talking with isn't listening anymore! One time when we were traveling, I was getting bags out of the car and someone asked me where we were coming from. The only thing I could think of was, "I'll have to tell you tomorrow." I couldn't remember that I lived in San Diego. I might be able to remember and say it tomorrow, but you can't have a very good conversation if you have to wait until tomorrow.

Mostly I try to joke about it, or if there is a family member nearby, I hope that they can interpret. I rely so much on Kathleen and the boys to decode what I want to say. This is somewhat of a rite in our family. When our son was in nursery school, he'd come home and Kathleen would ask him what he did that day. He would always say, "You tell me." So Kathleen would go through the litany of questions: "Did you play in the sandbox? Did you paint? Did you sing?" until he was downloaded and could respond. They all do the same with me now because sometimes I can only get out one letter of what I want to say—kind of like George Bush's T word for taxes.

Bill's resilient spirit was astounding. Many people periodically experience a word hovering on the tip of the tongue and yet just out of reach. The episode is frustrating, but usually, if the word does not materialize, we back off for a while until we can produce it later. For Bill, nearly every sentence evoked this experience. It required an extraordinary amount of effort just to speak. Each sentence built momentum, propelled by an urge to produce the words before they strayed off track and disappeared irretrievably into a neurological thicket. Occasionally, as if in a game of chance, Bill thrust out a word and a total surprise came forth. His startled expression would be overcome quickly by swells of his own laughter that released tension in recognition of the ridiculous. It was impossible not to laugh with Bill. Like Bea, his gift of escape from the tragic into the comedic was transcendent and lifted many people with whom he had contact out of the quagmire of disease and into an expansive vista of humor. But as much as we all savored this occasional reprieve, Bill's speech was clearly his most excruciating loss.

It is common for people with Alzheimer's to have difficulty with language over time. The disease initially strikes the memory center of the hippocampus and eventually moves across multiple regions of the brain responsible for language,

44

judgment, vision, movement, and behavior. But Bill's language function was hit particularly hard and early in the course of his disease and accounted for his severe aphasia. This pattern of impairment is not predictable and varies among people. Not only do people with Alzheimer's have distinct preexisting personalities and abilities, but—for no apparent reason—the disease may affect regions of the brain differently or more significantly in different people. Sometimes Alzheimer's is combined with other diseases such as Parkinson's, making the pattern even more complex. At the root of Bill's problem, however, lay that point of common ground among all individuals affected by Alzheimer's.

The Issue of Memory

I forget all of the time! Forgetfulness is half of my day! I think everyone forgets, but people with Alzheimer's have the problem more, and we get angrier more. I ask, "Why me? Why do I lose these things?"

My birthday is this week, and Kathleen bought me a desk. It's nice for me to have this new desk so that I can keep things in an orderly manner. But there was a lot of hullabaloo about getting the desk in and getting it assembled. There was one tool that I couldn't find. I just worked and worked to find it. I'd fume, "Where is that! Somebody took it!" And of course they didn't. It turned out that I picked this tool up and put it down in the garage and it was there all along. I just have to go and go and go until I find what I'm looking for. Kathleen usually finds things for me and that helps. I go to her first, and she usually knows where things are. Kathleen is very supportive. She's been a rock.

Unlike my amorphous first memories of Bill, my initial meeting with Kathleen is vivid. I was called into the nurse practitioner's office to introduce myself during Bill's first annual research center evaluation. While he was undergoing extensive neuropsychological testing, Kathleen was being interviewed to evaluate her perceptions of his impairments.

There was a gentle graciousness to Kathleen's demeanor that emanated through her reddened, tear-swollen face. Her soft English accent hinted at a reserve that could not be maintained in light of the overwhelming turbulence now affronting her life. Even the most stable of rocks can acquire vulnerable fissures under stress, and Kathleen had pressure coming from many directions. She was terribly displaced, not only by Bill's diagnosis but also by the return to the United States from their foreign service work and a move to San Diego where their son was the only mark of familiarity. Our task as a research center team was to pick up the pieces from a shattering diagnostic process and help Bill and Kathleen begin to reconstruct their lives.

Although Alzheimer's disease is a very challenging illness at any age, the lifestyle issues for those in midlife can be quite different from the issues for those in later life. After having worked for many years, couples are planning for retirement. Forced early retirement often prompts financial concerns and disrupts what is for many people a period of satisfaction in their careers. The hopes and dreams of "golden years" ahead are crushed by the onset of illness. Children may be at home or approaching independence at the same time a parent is facing a future of increased dependence. In a society that has discounted Alzheimer's as a disease of the elderly, younger families struggle to find a peer group and often find themselves included in a cohort of couples 10 to 40 years their senior. Bill and Kathleen maintained an intense interest in life and in learning as much as possible about Alzheimer's disease. Yet Alzheimer's also caused a separation from mainstream life. Bill and Kathleen were coping with challenges that the majority of their peers did not face, and finding social support was discouraging.

A New Opportunity

Recognizing a gap in services for couples like Bill and Kathleen, the staff at our research center developed an eight-week educational support group for couples interested in learning about living with Alzheimer's disease. Bill and Kathleen agreed to participate. Bill reflected on this experience in his manuscript.

I jumped at the opportunity. I had been diagnosed for two years but I had never known or even seen another patient! When the first day came, I was a bit nervous but I looked forward to meeting new friends that I could relate to. When we first glanced at the group, I thought we were in the wrong room because everyone else was at least ten or twenty years older than us. But that distinction began to melt away when we found that we were all in the same boat and in that sense, all the same age.

The real value of the group was that we were getting to know others—both patients and caregivers. I was also made aware of the different courses the disease takes with each person; I wasn't able to find my words, but others chattered along without a care. Sometimes they said the same thing over and over! Some of the group members seemed to have nothing wrong until you got to know them better.

Bill began to recognize that he was part of a community whose citizens, though united by Alzheimer's disease, remained distinct in their symptoms and coping styles. He was invigorated by the group experience.

After participants finished the eight-week series, we referred caregivers to the ongoing support groups available in the community. But there were no groups available for the people with Alzheimer's themselves. When Bill asked to see me during one of his appointments at the research center, his message was simple and direct—a result of his significant aphasia as well as his unwavering determination to be proactive in addressing the challenges of his disease. Each word came forth with intent and effort: "I want a support group for us—for patients. Not eight weeks—on and on." In his manuscript, Bill echoed his frustration: "I cannot understand why there are so many support groups for caregivers and none for the afflicted!"

I could not ignore the justice of Bill's request. Why limit a supportive experience to eight weeks when his disease is for life? Bill had an indomitable will to stay active and develop a social community. I sensed his impatience, as he was not sure how long he would be capable of making this request. I often hear from the caregiver, an essential voice in our care of the patient and his or her family. It is rare, however, that patients will make their needs known. I doubt that this is because they don't have any. Rather, they may passively, resentfully, or with some relief retreat behind the more intact verbal and functional abilities of the caregiver. Bill's request impressed upon me the importance of listening to the voice, however fragmented, of the person with Alzheimer's.

Maintaining Hope

An incentive underlying Bill's request for a support group was one common to people with Alzheimer's disease: the need for meaningful activity. There is often an imbalance between activities forfeited to the effects of Alzheimer's and new activities established in their wake. The experience of day-to-day satisfaction is subjective, and in optimal circumstances we have the opportunity to moderate our type and level of activity. But without options, one can become despondent and discouraged. Bill confirmed this problem: "I have some depression sometimes but I don't think it needs to be analyzed. I attribute it to doing the same thing over and over and over. Day to day, there isn't as much to look forward to."

In an attempt to address the activity needs of individuals with mild to moderate memory loss, the San Diego chapter of the Alzheimer's Association created a program called "The Morning Out Club." While I worked to develop an ongoing patient support group, Bill became a participant in this invaluable program.

At first, I was reluctant to go. I thought it would be like kindergarten. But from the beginning, I realized what a warm and caring group this was. The word *Alzheimer* is never mentioned, and we laugh a lot during the whole four hours. We start with coffee, but I take herbal tea. We discuss news of the day, do calisthenics, and then we work on a community service project. After our lunch, we have a sport or game, and then finish up with a word game. Once a month, instead of all of this we have an outing to a museum or some other place. It sounds benign, but the whole time there is planned so that we retain and expand our faculties. I'm delighted that such an opportunity exists.

I also work out, and I play tennis two times a week. I try to write on the computer as much as I can. And I watch a lot of TV. It's funny because Kathleen always hated television, but she knows that I need it now. It helps me because I can understand what I hear as long as the person doesn't speak too quickly.

I try to read, but it's very difficult. The lunacy of this gets to me. I forget what is on the page. I read along a line but then when I go to the next line, I can't find it. It takes so long to find the next line that I have forgotten what was on the previous line. It's so painstaking and slow. I can read little things, but I can't read a book anymore. Trying to read just tells me how bad I am.

It is important to have hope, though. What we're going to do in creating a support group will give us hope. It may not last long, or it may run out, but it's better than not doing it at all. The worst thing in Alzheimer's is that people don't get out. And there have to be better things to get them out. I don't want to vegetate. The Morning Out Club is supportive. At the Morning Out Club, no one ever talks about Alzheimer's and I like that. The support group would be a place to talk about it. People with Alzheimer's also need to get away—to see that there is life out there. So maybe we need a third group too! Alzheimer's is something to be scared of, but once you're scared you know that you're not going to die right away, or even that year.

As I walked through Bill and Kathleen's garden gate two and a half years later, I reflected on the time since my last interview with Bill. The entry garden was different now, and I paused for a few moments to absorb the space. White fortnight lilies and purple agapanthas were in bloom, and the sound of trickling water beckoned from some unseen spot. A small stone shrine emerged from the blossoms.

This time, when Bill and Kathleen greeted me at the door, I looked on the faces of two people who had become a part of my weekly rhythms at the research center. Bill's requested support group had materialized, and for the past 28 months, he had rarely missed a week. We were moving through time together in a unique way that I experienced only with members of this support group. Each person was becoming infused into my life like gradations of color that begin in the faded undertones of a wash on canvas and progress with each application to the rich dimensions of a painting. Each of the losses they forfeited to Alzheimer's disease, while subtracting from aspects of their life, laid yet another coat of pigment on the canvas and prompted me to deepen my vision to see each person's ever-changing portrait.

Bill's losses continued to be most prominent in the area of language. This time, Kathleen joined us at the round dining room table for the interview. Kathleen was an invaluable resource for Bill, having learned to decipher his thoughts from the lead of one or two key words or a pantomimed gesture. He was reassured by her presence, as one would feel grateful to an interpreter when communicating in a foreign language.

Once again, Bill had centered his manuscript on the table. Although writing the manuscript had been a therapeutic experience, the presence of it now elicited painful feelings. While the text had once been molded around Bill like a familiar pair of shoes, the contours of his own self had transformed so that the shoes no longer fit. Bill could no longer write or type. He had recently dismantled his whole computer system because its presence was a painful reminder of his loss. "I can't believe I wrote this," he stammered in reference to the manuscript. "It makes me sad. Where did all of that ability go?"

Bill's defeated expression was painful to witness. Sometimes the only way I can cope with seeing people I care deeply about progress in this disease is to focus on the abilities that remain and pull out the dimensions of their personality that still exist but become expressed differently as a result of the bizarre language of the disease. Yet it is disingenuous always to be optimistic. There must be room for grief and for an acknowledgment of losses fundamental to one's very sense of self. In the silence that followed Bill's comment, we sat in a sad yet sacred space—a realm where the comfort of relationship absorbed the loss and softened it through the cushion of human connection and trust. Given how deeply Bill mourned the absence of his previous abilities and adult competencies, I was surprised by his next comment.

I'm becoming more childlike now. I enjoy the things children enjoy. I don't have the same responsibilities. I can do what I want. I really am a child.

Sometimes the simple clarity of Bill's sparse commentary was startling. His comment was without the disdain that I frequently hear from professionals or caregivers who speak of the tragedy of a regression to the helpless, incapacitated state of childhood. Indeed, this increased dependency places a large responsibility on a care provider, much as caring for a dependent child places that responsibility on a parent. And unlike the unfolding joys of witnessing a child's development, an adult's regression is predominantly linked with loss, despair, and humiliation.

But Bill's comment did not evoke tragedy; rather, he recognized an ironic innocence in his simpler life. Surely, if given a choice, he would be cured of Alzheimer's disease. But without this option, he was editing the verbose narrative of his life into haiku—poems whose paucity of words revealed not so much a poverty of language as the richness of each given word. Bill treasured life. And although Bill and Kathleen both continued to grieve his changing condition, there were times when his childlike simplicity could be restorative.

"We'll go on a walk," Kathleen reflected, "and he'll be so excited about a rabbit or some other animal we may see. On one of our walks, there were a lot of snails on the path, and people crushed them as they walked. Bill was stopping and picking up the large ones and throwing them in the shrubs, trying to save their lives. These are the precepts of Buddhism—the reverence for life—that have stuck with me, and with Bill, too. Before Alzheimer's, he probably wouldn't have stopped; there was always something else going on. So, you just slow down and appreciate the small things."

Both Bill and Kathleen were stressed by Alzheimer's disease. But as his life simplified, her responsibilities multiplied. I was moved by Kathleen's genuine ability to find inspiration in such a challenge.

Since our last visit, Bill had found encouragement in new arenas. His quest for meaningful activity had been fruitful, *and unlike a few years ago, he now felt he had things to look forward to every day: "I have all types of things to do. I eat well. I walk a lot. I'm at the support group every week except for the times we go away. I love the group. It's joyous. It can't be bottled."*

Bill and Kathleen had found a rich community in the support group. Although we did not facilitate a formal meeting for family members, they often socialized while their loved ones were in group. Despite his severe language disturbance, Bill was an active member in the group. He often brought in Alzheimer's-related news clippings or questions he'd asked Kathleen to write down for him, and his animated gestures expressed his response to group process and discussion. Occasionally we read from his manuscript and, ritually, from a collection of poems he'd written for Kathleen over the course of their marriage. The verses now served as a closing passage for each group session and a tribute to the mastery of words he could no longer command. In spite of his profound appreciation for the group, I wondered if Bill felt any adverse effects from his participation, and I mentioned a prevailing concern that it could be depressing for people with Alzheimer's to witness the troubles and progressive decline of others with the same disease.

"No! No! Who says that?" Bill's face contorted into an expression of irritated defiance. "They have groups for cancer. How could anybody say that?"

Indeed, Bill had witnessed a great deal in the past two years of the group. Chuck, his tennis partner and fellow group member, had died suddenly of a congenital heart defect. Although Bill and Kathleen were in England at the time, many of the group members and their spouses attended the memorial and mourned the loss of this valued friend. Other group members showed increasing signs of impairment, and none were able to drive. Bill had voluntarily given up driving, but still clung to his daily solitary walks in the neighborhood. He knew that his willful hold on some semblance of independence placed him at risk for a frightening dimension of Alzheimer's disease.

"Poster Boy"

The National Alzheimer's Association estimates that 60 percent of people diagnosed with Alzheimer's wander or become lost at some point in the course of the disease. People may become disoriented in previously familiar places, or the behavior can be more persistent—a restlessness that is motivated by changes in the brain, anxious feelings, the physical comfort of movement, or a search for familiar ground in an increasingly alien world.

Bill knew that although he was currently able to navigate his well-known walking route, his disease could ultimately, without forewarning, render the known unknown. Episodes of confusion can be extremely frightening for people with Alzheimer's disease, and Bill had heard accounts from support group members who had become completely disoriented while close to home.

In response to this concern, the Alzheimer's Association established "Safe Return," a program designed to locate people with Alzheimer's disease and return them to safety. A specialized identification bracelet links the person with a nationwide network designed to help reunite the lost individual with his or her family and home. Some people are reluctant to wear the bracelet for fear of stigma. Others deny any risk of becoming lost. But Bill's interpretation of the program characterized a more positive adaptation to his impairment. He recognized it as a safeguard to his treasured sense of independence. He was delighted to comply when the San Diego chapter asked him to be a "poster boy" for the program. The poster depicts his warm smile and his message to the public: "I like to take long walks and Kathleen was always worried that I'd get lost. Now we both have peace of mind. It also gives me more confidence when I travel. She can always find me."

Adjustments and Adaptations

As Bill's participation in Safe Return illustrates, Bill and Kathleen confronted each loss strategically. Since our last interview, Bill had relinquished his exasperating attempts to read and now subscribed to books on tape. Originally established for the blind, this program provided him with a specially equipped tape player and library privileges for the enormous array of literature recorded for the listener. Bill's memory impairment made it difficult for him to follow the narration of a long, complex novel. Instead, he acquired poetry, weekly news journals, and short stories. He was very pleased with the program because it allowed him to accommodate his persistent intellectual curiosity and skirt one more obstacle presented by Alzheimer's disease: "You just sit there and it all comes to you. I don't have to listen to all of it, but when I want to focus on something, I can do it. All in all, I'm adapting well."

In the years since Bill's first evaluation at our research center, our interactions have fostered an extraordinary kind of dialogue: our communication has evolved while Bill's very capacity to express verbal or written language has simultaneously and dramatically declined. Throughout our interactions, I have grown increasingly sensitive to the ways in which Bill's life and method of coping with Alzheimer's disease convey many important messages. He has prompted me to reflect on how few words we need to express the essential.

We often organize our lives via a succession of events—a time line of experiences that build on one another and lead us to frame a logical sequence to our lives. Events become detailed narrative, and our life story unfolds. We reflect on the past; we plan and hope for the future. But for Bill, the past is elusive, the future uncertain. A relationship with him can be demanding— but also deeply inspiring. Like a Zen master, he exacts my conscious attention to each present moment and renews my appreciation for the unpredictable and spontaneous dimensions of life. Despite significant spells of despair, he remains curious about life and about the discoveries available in each day. He delights in the offerings of the natural world, and by advocating for a support group, he has been instrumental in conceiving new opportunities to maintain connections to a social world.

Just as the text of his manuscript enhances this chapter, Bill's contributions to the text of my own life are considerable. None of us really writes our life story alone. By invitation or by intrusion, other influences dictate some of the script. In turn, we respond, we edit, and we invent. Through my interactions with Bill, pieces of my own life narrative are infused with his influence. He is not simply a character in sections of my autobiography; he is a coauthor. I am a bit different for knowing him.

Surely, in this place and time, introduced to each other by Alzheimer's disease, Bill and I are compelled to collaborate on a work in progress. As far as we know, his disease will progress. But it is also true that our understanding of Alzheimer's disease will advance and lead us to new treatments and interventions to address the multiple dimensions of this disease. Bill will want to participate in and contribute to these developments in whatever capacity he is able. Robbed of his written and verbal language, but ever striving to express himself, he now communicates through his actions: advocating for Safe Return, saving a snail's life, contributing to research, consoling a friend in the support group. With limited words, Bill's voice remains resolute.

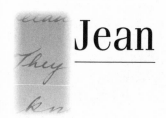

Jean

"I'm almost 71 and I'm not amazed that people die. So it isn't the death. It's the loss of oneself while you're still alive."

My first contact with Jean was by phone. She was referred to me by her psychologist, who felt she might benefit from the Alzheimer's support group I was developing as a result of Bill's prompting. Widowed just under a year previously, Jean was experiencing the double blow of the death of her husband of 47 years and the recognition, via a work-up for memory loss, that she was probably in the early stages of Alzheimer's disease. Her newly imposed solitary status had recently converged with her developing memory loss in a terrifying episode of becoming lost while out alone doing errands. Although the idea of a support group appealed to her, the logistics of transporting herself to the group felt impossibly complex and frightening.

I suggested she take a taxi to our first group meeting. There was an amused yet resigned quality to Jean's response that indicated she knew this piece of dialogue all too well. "It all sounds fine now—very reasonable," she began, "But as the time gets closer and closer, I'll freeze up. What if the driver doesn't know where to go? I won't know him. He might get us lost."

I had to acknowledge the truth in Jean's predicament. Although it was unlikely that the taxi driver would become lost on the well-defined route to our building, it was frightening to entrust herself to a stranger. Life mandates a degree of ongoing trust in ourselves and in others to function day to day. Hopefully, we acquire the competence necessary to negotiate interactions and to respond to unforeseen circumstances. Yet for Jean, that confidence, so fundamental in life, was clearly eroding.

Early in our conversation, a quality in Jean's voice engaged my attention. She seemed to speak inwardly and outwardly at the same time. Her internal world projected outward through a voice that transmitted words while concurrently conveying reflection. It was a dimension of Jean that I would come to know in greater depth: she experienced her life with all of her feelings, thoughts, and behaviors, while also maintaining a capacity to observe them. Jean was both on stage and in the audience—acting in her own life script while watching other forces take an increasingly dominant role in the directing.

I quickly became aware of my protective feelings toward her—a temptation to become the director. But the impulse was tempered by my respect for her insight. Her anxiety was palpable, not through exaggerated outward expression but in the gripping way that she described her internal immobilization. Her fear of unmanageable consequences resulted in painful episodes of inertia, and it was this frozen quality that was paradoxical. It seemed contradictory that her bright, perceptive mind should speak to her in a manner that evoked such paralysis.

"Your voice is so soothing," Jean told me in that first conversation. Oddly, even in her anxious and uncertain state, she had a calming effect on me as well. Rather than pushing against the formidable transportation obstacle, I relinquished the urge to remedy her predicament. Jean was inviting me into her life with both caution and gratitude. Any more coaxing on my part would have disregarded her vulnerability and

furthered her isolation. Instead of requiring that she find her way to our support group meeting, I invited myself to her home. Jean was noticeably relieved. We set a date, and she gave me directions.

On the scheduled day, fifteen minutes after our appointed time, I arrived at her front door. Although I'd had no trouble finding her condominium complex, negotiating the maze of pathways to her unit proved daunting. Each path appeared to circle back upon itself, and the sequencing of addresses was baffling. Pure luck enabled me to stumble upon the cluster of condominiums that contained her residence. The housing complex was certainly incompatible with my sense of direction, and I became aware of an ironic empathy for the disorientation common to people with Alzheimer's disease.

"I got lost," I proclaimed as Jean opened the door.

Looking both concerned and humored, she laughed wryly. "Well, we have something in common."

Jean's long, straight white hair was pulled back in a ponytail, and her face conveyed a smooth, soft radiance that seemed contrary to any history of anxiety. Once inside, her home spoke of psychological and artistic inquiry. In the innocuous environment of her condominium complex, she and her late husband had established a small pocket of identity together. Subdued in color, the artwork and small ethnic sculptures that filled the room had a sparse yet vital quality. It felt as if each piece procured over the years could tell the story of a particular time or experience in their life together. I was to learn later that Jean herself was an artist. She also loved the blues.

Jean made us cups of tea, and we sat down at her dining room table. I felt the immediate ease of two people who were at home with the explorative travel of good conversation. There was little to sift through to get to the heart of life, and as I asked my initial questions, Jean entered a reflective realm where she both experienced and analyzed her complex world— a world in flux since the introduction of Alzheimer's.

The Onset

I guess I've had some memory loss off and on for years. I'd think, "Well, I'm almost 70." It wasn't anything that alarmed me, and most of the things that I forgot really weren't that important. If I really needed the information, I'd be able to track it down. It's when words began to slip away—when I couldn't finish a sentence because I'd lost the main part. When I'm trying to express something to someone and I get a third of the way in and I haven't got a vital word, that's annoying!

The memory is a loss that is very painful because you know you have something in there about whatever the issue is and you can't get at it. There is embarrassment when I want to say "ocean" and I can't think of the word. It depends on how comfortable I am with the person I'm talking with. Then I can ask, "What's that big water thing?" and they'll guess, "The ocean?" Then I say, "Oh yeah," and I can go on with my story. There is the issue of safety. If I'm in a group of women with whom I am comfortable, I'm not as likely to have a problem. But if it is somebody that I don't know, then I feel like I'm practically scaring them. With strangers, they don't know why you don't know what you are saying and they get confused and disturbed. With people with whom I'm not familiar, it's more likely to happen. It's stress.

I notice on the phone very often I can't think of my address. I have a little piece of card on my desk with my name and address and all of that stuff. But darn it if they don't always come up with another question that isn't covered in my papers! The list gets bigger and bigger. The more rattled I get, the less I can pull out. So when somebody asks, "Well, what part of the street are you on?" I'm just finished! I can't do anything else because of the stress.

It isn't always terribly painful. I'm just as likely to talk to people as I was in the past because I forget that this can happen. I've been able to speak for so long that I don't usually

60

think, "Oh gee, I hope I can talk!" It's not like I'm on TV. Now that would be difficult. But every once in a while there is something that I really want to talk about, and I just back off because I haven't got the stuff to do it with. It's a loss, a really big loss. It could be something very trivial, or it could be something very significant to what we are talking about, and they are equally disturbing. It's the loss of the ability to express yourself.

The onset of Alzheimer's disease is usually marked by changes in both short-term memory and language fluency. People commonly excuse the memory loss and attribute it to aging, stress, disorganization, or myriad other reasons. But language changes can be difficult to disguise from oneself or from others. Jean's language disturbance was subtle and less easily discerned than Bill's much more profound aphasia. I was engaged by her ability to articulate complex experiences and feelings. As the listener, however, I was not privy to the concentration that was required for her to convey her message. The pauses or occasional grasping for a word could have been interpreted as the process we all go through when we try to form thought into language—when we seek the right descriptive word or phrase. Yet for Jean, it was not a matter of enhancing her verbal expression. The occasional broken rhythm revealed a more discordant process that interrupted her communication instead of embellishing it.

Jean's descriptive account of her language difficulties made me aware of the private dimensions of Alzheimer's disease—of how much a person may be experiencing that never reaches our ears. We tend to respond to the outcome and not inquire about the process. As Jean began to discuss the medical evaluation of her early symptoms, it again became clear that her response to the diagnosis was accompanied by an inner process of associations that, without inquiry, would have remained unexpressed.

The Diagnosis

I went to a lot of doctors trying to find out what was wrong with me. I didn't even really know what the problem was, but I knew something was amiss. Somebody finally said that they thought it could be the beginnings of Alzheimer's. The diagnosis was very fuzzy. I felt pretty bad. I remembered one of my dearest friends I'd made when my kids were little. I got a card from her one time saying, "I'm losing my marbles." And I thought, "Oh, it's Christmas time and she's rushing too much." I didn't really know what she meant. We were back east then, and she had moved to California. When we came out here to live, I had an opportunity to go see her. A friend told me she was in a facility with Alzheimer's. When I saw her, I was not prepared for the damage that had been done to her. She recognized me, tried to speak to me, and did say a few words. She was very excited that I was there. I cried for three days. So that was my thought when I first heard my diagnosis. That was my image of Alzheimer's.

My biggest concern is that I'll eventually be unable to care for myself and be in a facility. I used to work as a secretary at Mount Sinai Hospital in New York, so I got to learn a lot about nurses. I learned that some were such kind people and some were awful. If you don't have control over yourself, it's better to have a kind and caring person. But that's not the real issue. It's that you aren't you anymore. You can't do the things that are important to you. And I know that happens to all kinds of people, not just with Alzheimer's. But when I think of my friend Lois, I think about the way I knew her initially and then how I saw her in the end. Part of her was still Lois, though.

A diagnosis of Alzheimer's disease in its early stage often leads people to images of late-stage Alzheimer's. Our society and the media frequently focus on the tragic devastation of the disease. Ask most people their first image of a person with

Alzheimer's, and you will usually hear a description of a severely impaired person in a nursing home. As we diagnose this disease earlier in its progression, there is a growing population of people who are only mildly impaired by the disease at diagnosis and, indeed, may have many years of relatively mild impairment ahead. It is no wonder that we often see denial as a response to the diagnosis. Many individuals' conditions do not match their own perceptions of the disease. Hence the common experience of "That can't be me." Friends and family are not excluded from this response and occasionally maintain denial even after the person with Alzheimer's has begun to accept the diagnosis.

Shifts and Changes

While facing her diagnosis, Jean began to notice the ways she was affected by Alzheimer's. Some of the changes were of little concern; others posed greater significance and indicated shifts in her feelings of security and independence.

I know there are areas that are slipping, but I don't quite know offhand what they are. The other day I was going to put some laundry in the laundry room and I had to get the keys to open the door and get the soap and all that stuff you have to schlep with you. I found it very difficult to get it all together. But it was of no significance that day. No one was waiting for me, and I didn't really care if it took me another 20 minutes to get it together. I wasn't functioning well that particular day, and I couldn't even do the little stuff. I recognized that and accepted it and thought, "Well, just keep going till you get it all together and then you'll go and do the laundry." It wasn't a shame because nobody else was involved.

There are some scary things, like not remembering my name. I remember the first time that it happened: I was in a bank and I had to tell my name. I always thought I had enough trouble telling what my mother's name was! But I couldn't think of my own name and I said, "Oh, never mind. I'll come back." That was really difficult. It's like something hit me hard. I'd already indicated that I was going to give this important material and all of a sudden I wasn't going to. So it's not being able to meet their expectations, or even my own. I certainly don't feel like I have full control. And some days I don't feel like I have any!

There is an unpredictable quality to living with Alzheimer's disease. The daily tasks of living can become interrupted by a confrontation with memory loss like a sudden power outage. Sometimes one can ride it out. At other times, the darkness is frightening and a reminder of just how much we depend on smooth circuitry and ready access to our intellect. Jean was able to be fairly self-reliant in familiar surroundings, but when she was away from home, her memory was not reliable. This resulted in one of the more frightening episodes she encountered in her experience of Alzheimer's.

I wasn't having trouble with finding places maybe six months ago. Recently, though, I was driving in a familiar area but I had gone farther than I usually go. There were buildings on either side, but nothing seemed to be open. It was like a nightmare. So I went back and forth and back and forth, and I was afraid to move more because I really didn't know where I was. I finally saw a door halfway open on somebody's building, so I asked for directions. The women who helped me out—they were saints. They were the sweetest, loveliest women. I'll never forget them. But I'll never find them again, either! They'll never know how grateful I am. In fact, a friend offered to help me find them, but I wouldn't be able to. The women carefully wrote perfect directions for me to get back on the freeway. I was OK when I knew what to do next. But the fear at the time was that I was lost forever. There was no way to make the fear go away. Now I can only drive in certain areas I am really familiar with, so that limits me enormously. To be able to drive means that you can go and do what you want to. It's not like I'm a big shopper or anything. But it's a loss. That's all I can say. It's just a loss.

One thing I've noticed is this pulling back. I desperately want to go somewhere. I must go. And then I start pulling away and pulling away. There are too many things that are too scary. If I take myself away from the very thing I want to do, it helps me because I avoid the fear of having to get there. I'm afraid of being in the car and not knowing where I'm going. It's a fear of being stranded. The fear is so strong I can't describe it. It's like I'm saving my life. I don't go anywhere when that happens. So I wound myself by trying to protect myself.

I'll have a friend drive me home and she won't be familiar with the area and she'll ask, "So, do I go right here?" And I'll chuckle, "Do you think I know? You've got to be kidding!" But eventually something looks familiar. And at the time I say to myself, "Jean, even if you don't know where you are, you're not going to die from this." But I always feel a bit of life being taken away when I have to back off. If you add it all up, you get a very thin path to use.

Although she recalls many incidents of fear and loss since living with Alzheimer's disease, Jean recounts with equal vividness those episodes when someone has shown goodwill—the drawing of a map, a listening ear—times when by an individual's words or actions, the world momentarily felt safer and less alien. Listening to her, I felt a sharpened awareness of the impact we can have on one another as friends, as family, or as absolute strangers. Like Jean, I have particular memories, some painful, some deeply encouraging, that create a blueprint for the foundation of my relationship to others. Some memories erode over time; some remain like cornerstones that validate my reality or inspire the lessons I will draw upon to face a given circumstance. And new memories are made in daily living where, by some neurological mystery, I will filter out so much and hold on to seemingly so little. What remains is what becomes real in my world. With Alzheimer's disease, what remains may vary from moment to moment. What is real can become nebulous.

And in this changing reality, there is a feeling of losing both control and independence. Jean experienced a shift not only in her capacity to navigate the familiar surroundings of a neighborhood but also in her knowledge of the well-worn territory of her own self. Jean suffered a frightening loss of orientation to aspects of her identity, which was projected outward into an overwhelming fear of getting lost in the world around her. She recognized that with the onset of Alzheimer's, she was far less self-reliant. And as we spoke, it became apparent that without her husband, Jean felt alone and uncertain on whom she could depend.

Confronting Loneliness

My husband's death a year and a half ago was a big kick in my memory. When Harry was here, half the time when I was having problems, he could help me through and get me back on track. He was somebody I lived with. I have had to get used to the loss of somebody who understands me. It would be going on 49 years, or close to that.

I often think of how Harry would be responding to this. He'd deny it. "You don't have Alzheimer's." In his mind, I was never sick, never had a cold, so it could never happen. I don't think his response would have had much of an impact on me. I would notice how he was handling it, but it wouldn't change how I feel about what is happening to me. He certainly never mentioned Alzheimer's. I think he was afraid of a lot of things happening, including dying, and he just wouldn't talk about it. I talk about it. It's something we all think about from time to time.

I feel tremendous loneliness. I was very active in the women's movement, and all these women I know are still working. They're much younger than me, and they're all busy. So it's not as though I can call upon them anytime to go to a movie. It just isn't that way. But I'm in two women's groups. One of them is tonight and I have to figure out how to get there. I started participating in women's groups when I lived in New York. This was before NOW *(National Organization of Women)* and this was the radical feminist movement. Consciousness raising was the issue, and we just went on and on with it. So these women became my family. Age didn't mean anything. Nothing meant anything except our belief in ourselves. So, I stayed with consciousness raising. It was always something that worked very well. The whole feeling was really sisterhood. I've been involved in the groups with NOW out here. So every once in a while somebody will come up to me and say, "Oh, do you remember me?" And I'll think, "Oh dear, she was in one of those groups," and I don't have any idea who she is. But they still remember, 20-some years later, how significant it was.

I've told them about Alzheimer's disease. They are very quiet. They don't know what to say. I don't know what to say. Sometimes I talk about it, but it's the kind of group that doesn't tell each other what to do, so there is not a whole hell of a lot that they could say. But they are saddened and moved, and you just have to swallow it. I think they understand because I'm telling them why it's so hard and the impact that it has. So, they know what I'm saying. I think that we tend not to respond a great deal. We listen. I don't expect them to respond any more than I could have responded two years ago. I know that they grieve the fact that I have this, but there is nothing to be said. You do what you have to do. So I don't expect more than to really have an opportunity to say what is going on and to express how I feel about it.

Jean's message reminded me of the importance of being heard. As she spoke of her experience in consciousness-raising groups, I reflected on my own similar experiences with radical feminist groups one generation later. In courageously examining her roles and identity as a female, Jean had participated in a demolition of the boundaries traditionally surrounding women in an evolving Western society. The radical feminist premise is that "the personal is political." One cannot view personal problems as separate from the influences of the social and political context of the dominant society. Our common experiences in these groups taught us not only to speak of our personal concerns but also to motivate society to understand and alter its impact on these same problems.

Through the process of our conversation, I became aware that a dimension of this lesson continued. In her willingness to speak about her experience of living with the disease, Jean was continuing to deconstruct boundaries—those barriers of silence, misunderstanding, and neurological degeneration that can keep us from truly understanding the perspective of the person with Alzheimer's. If we are to affect positively the personal experience of those living with Alzheimer's disease, we must also address the larger medical and social perspectives: our health care policies; our medical treatment and research funding; our care of families; our images of people with Alzheimer's. The personal becomes political.

Accepting Help

The experience of being alone created new challenges for Jean. She vacillated, as many would, between her need for assistance and her desire for independence. As one whose profession is based on understanding, addressing, and attempting to meet human needs, I listened intently as Jean again referred to her need to be heard.

There are a lot of people who help you in the beginning. That's their job. There are people who very early in this process said you must get someone to do your checkbook. I wasn't at that point yet. It was very insulting to me to be told, "Never mind what you think. This is what I think you should do." There is a lot of that attitude in these well-intentioned people. I fought like hell during every single step in getting help. I'd argue, "I don't need it. I don't need it yet. I don't want it now," and so forth. Then eventually I think, "I really do need it now," and I do it. I don't know at what point this happens. It's something magical, I guess. Yesterday it was out of the question, and today it's, "Ah, just do it. It won't hurt you. What's the problem?" Then I'm always glad I did do it after all. It's a slow process, I guess. With the checkbook, I knew when it was time to give it up. Let somebody else do it. This charming woman is just great with my checkbook! I can work it out, but it takes a lot of imagination to get it. But the decision had to be at the right time, and people were insisting at the very beginning when I didn't feel I had to do anything. But they call you back and maybe it will work next time. Sometimes I'm just not ready. I understand that it's very important to do, but I'm not there yet. I guess I have to wait until I'm not pushed because if I'm pushed, then I'm never sure if it's what I really want to do.

I understood Jean's stance. Who among us leaps at the opportunity to be told how to live our lives? I usually need time to reflect on my circumstances—to weigh the effects of various courses of action. Input is valuable when requested. But ultimately, I want to be the primary participant in the decisions that concern my life. Yet, I understood equally well the intentions of the well-meaning souls who insisted early on that Jean seek help. A reasonable perspective is that she was taking too long to examine her circumstances, and intervention was warranted. There are times when the decisions that we *do or do not make about our personal lives have a detrimental impact on ourselves or on others.*

An ingrained value in the social work profession is "client self-determination." The premise is that people have the right to make choices about the acceptance or refusal of help and that we all have the right not to have choices regarding the conduct of our lives imposed upon us. This value stems from the constitutional rights of privacy and free will. The only legal and professional exception is when an individual's behavior poses a grave danger to himself or herself or to others. This issue is complicated when intellectual functioning is in question. The questions arise: How grave is the threat to one's own or another's safety, and how well does this person understand his or her circumstances and the consequences of his or her decisions? If no family member is available to assume responsibility, and a gravely disabled person refuses help, a court may have to appoint a guardian to make decisions on the person's behalf.

Jean has two sons, but both live on the East Coast. Although initially defensive, she eventually recognized her need for assistance and consented to having a local social service agency help manage her finances. The checkbook is a menace to people with Alzheimer's disease. It becomes increasingly difficult to record and calculate numbers, pay bills, and balance accounts. It is usually necessary to relinquish this task to a trusted assistant. But the manner in which this is handled will vary from person to person. Embodied in that "magic moment" when Jean turned over her checkbook was an acknowledgment of her own changing ability to negotiate this piece of life. Yet she was able to transform the power of the checkbook from something that symbolized her autonomy and competency into a tedious and frustrating chore that could be better accomplished by someone else.

I appreciated Jean's fighting spirit. She was able to direct much of her outrage at the disease itself and thereby lessen *the risk of displacing that anger onto her relationships with others. Although people could be insulting in their approach to her, it was the fact that the disease even existed that was the ultimate insult.*

In Search of Justice

There shouldn't be such a thing as Alzheimer's! I never thought, "Why me? Somebody else should have it instead of me." I don't think it should be them instead of me. There isn't any chart that says this is not all right for me but it is for her. There is no sense to it at all to begin with. And if there were somebody who could control these things, then they ought to do a better job. Sloppy, sloppy work! *(Jean directs her scold toward the heavens.)* I think I've probably used the "there's nothing you can do" attitude. I don't know why this is happening. It doesn't feel like it has a meaning. It just happens. It happens like I hurt my knee. I don't like that either. But it wasn't because I kicked the cat or something.

A friend of mine gave me a book on a woman who was writing about her Alzheimer's and how gloriously, wonderfully she was handling her disease. I hated that book. I wouldn't consider behaving as well as she did. I guess there are people out there handling this marvelously. But it just isn't me. I want to cry and whine and kick! Not as ladylike, I must admit. There is a little fighting, but mostly it's whining. Its mostly, "Damn it! This is terrible! This shouldn't happen!" Then of course other words come in. There is my goody-goody voice going on all the time that says, "There are a lot of things that are terrible, Jean. A lot of people are very sick with terrible things and have real physical pain." You know, this whole story. And then I think, "Stop it! That has nothing to do with me. I have Alzheimer's and I don't like it!" I'm angry. I'm angry that anybody has it. But I have that voice that says be nice and you'll feel better. I'm almost 71 and I'm not amazed that people die. So it isn't the death. It's the loss of oneself while you're still alive. There are other ways to go through that process, too. There are all sorts of illnesses where one could have no control over themselves, and I would cry if I had one of them, too.

There were times in my conversation with Jean when I was aware of a poignant paradox. Ours was a new relationship. Throughout our dialogue, I was learning about Jean through her description of her history, feelings, thoughts, and behaviors. A self was unfolding and taking form. I had no historical relationship with who she was, and I had no expectations of who she was supposed to be. Yet in my mind she was becoming known to me at the same time that she feared becoming unknown to herself. Like most people, Jean's sense of identity was influenced by her ability to do the things that were important to her, and it was this dimension of her self-concept that was most affected by her early-stage disease. She was losing the ability to control her speech, to navigate with a sense of direction, and to track all of the steps necessary to complete a complex task.

But how do we gauge the complete loss of self? Even when severely impaired, no one person with Alzheimer's disease is exactly like another. I reminded Jean of her friend Lois in the nursing home—and how part of her seemed the same. Personal identity undergoes profound changes with the progression of the disease, forcing people with Alzheimer's, their families, and friends to adjust their expectations continually. Yet the impairments in thinking brought about by Alzheimer's often affect patients' awareness of these very changes. Insight varies considerably from person to person. Some, like Bea, understand and can communicate the nature of their illness well into the advanced stages of disease. Others have only episodes of clarity when a particular sparkle of previous personality shines through with startling brilliance. And some are so transformed by the disease that they seem considerably less aware of their condition.

A Message to Others

Jean monitored herself both emotionally and behaviorally. She sought to understand Alzheimer's, to learn to cope with it, and to teach others in the process.

My advice to people with Alzheimer's is don't have it! But if you do, get as much help as you can to learn about what you can do, what it might be, and what not to do. Just have as clear a picture as you can about all of the possibilities. Try to prepare yourself about them and about what you can do, if anything. I'm telling other people to do this, and I'm not so sure I can do it myself.

To family members, I say, "Don't baby me, and don't pretend it isn't there." So far I haven't had any problems with family members. To friends, I say, "Let me talk about it sometimes. Whether I'm angry or sad, just listen." I'm sure almost every doctor I went to knew that Alzheimer's disease was a part of this, and they didn't want to deal with it. I'm very angry about that. In some cases they said, "Well, you're never really sure about these things." I can understand that you could be cautious in some cases, but they need to give the person some idea of what they might have instead of protecting themselves. You have to keep pushing until you get someone who will tell you what you need to know. Try to get someone to help you through and think about it. We don't have a lot of people in our lives who know what we're talking about.

A year and a half later, I revisited Jean in her home. This time, I found it with ease. The house was busy with visitors from New York—old friends who came on vacation and stayed with Jean once or twice a year. The flurry of activity filled the home with a kind of background noise, and I wondered if my own visit was poorly timed. But forewarned of our meeting, the friends left for a few hours, leaving Jean and me seated once again at the dining room table.

It was striking how Jean could refocus so quickly. Excess stimuli often exacerbate the memory and confusion problems of people with Alzheimer's disease. But once the visitors left, it was as if they had never been there. The immediate moment was all that existed as we began our conversation. It was difficult to determine if her short-term memory loss had caused Jean to forget the hum of activity that had existed only moments before. Or was I witnessing once again her ability to access her feelings and self-awareness—living in the hustle and bustle, with the capacity to step back into observational retreat?

We reviewed the events of her life over the past year and a half. Jean had become a founding member of our research center's Alzheimer's support group, and through the friendship of another group member, she was provided with weekly transportation to the meetings. After an experience with an incompatible roommate, Jean was again living alone. She continued to struggle with feelings of loneliness and fear but was ambivalent about trying to find another roommate. Although residing in a care facility had once been the symbol of a frightening decline, Jean now spoke of the sense of security and possible community that a facility might afford.

I think living alone at this point now is something that is just plain old. I want some interaction with people. I think about moving, but where to? Is it worth the move, and will I be making another move that is more permanent? My son and I looked at a few retirement facilities here and a few places back East where he lives. There was a wonderful one here and a wonderful one there. I guess when the time comes, I'll have to make a decision. That would be if something happens like I break an ankle or something. Then I'll know that I can't do it on my own.

Although she still experienced episodes of considerable anxiety, Jean also described an adaptation to her condition. The disbelief and anger about her losses had settled into begrudging tolerance. Memory loss was familiar to her. It no longer commanded her full attention or scrutiny. Rather than fearing it, she found ways to skirt around the obstacle—to avoid confrontation and undue challenges. Jean modified her expectations of life to include a degree of unpredictability and was no longer startled by her condition.

I used to get more upset about it. *I* didn't forget things. But now, sometimes there is a connection and sometimes there isn't. I don't get terribly upset about it. Either it will come back to me or it's not important. Or it's important, and I still don't have it. Whatever it was, I'll either think of it again or I won't. Neither one will kill me.

I guess in the beginning, I was kind of scared by all of the losses. I think I've learned to just drop it. I tell myself, "You're not going to do that. And you never learned to skate on ice either, so just give it up!" Even if I don't want to do something because I'm scared, that's just as important. A lot of people won't allow themselves to be overwhelmed by something. But for me I think, "Don't do this—it will only make it worse." I'm scared for a good reason. Maybe not somebody else's good reason, but my own. When I was young, I always feared I wouldn't be able to do what everyone else could do. But none of that is as important as it used to be.

With the passage of time, Jean had renounced some of the fight. This did not result in apathy or indifference. Rather, she seemed to feel that her current circumstances no longer warranted the protest so pervasive in our first meeting. As a weekly participant of the support group, she was attentive but fairly quiet. When other group members engaged her, Jean's responses were thoughtful, direct, and often accompanied by her candid humor. Although she was not inclined to paint a rosy picture of her circumstances, her comments did not convey self-pity.

I don't think I whine at all anymore. The support group has been a wonderful experience. They are all such good people. They're trying to share. And nobody is whining about it. That's a big step right there. What's to whine? I can't point to something and say I used to play the guitar and now I can't. It isn't like that. Maybe in the beginning—although I can't really be sure what I remember—anything like that was upsetting. If I remembered someone's telephone number but not the address, I'd be really angry with myself. But now if I've got anything that I can use, I'll use it and if not, I haven't got it! I think that what I taught myself to do without thinking about it is if something is getting difficult, I just push it away and go about my business. I think this also has a cost involved in that I've had to give up some things that I cared about. But it's not that bad. Music is just as wonderful to me as ever. I think a good part of myself is the same.

Shortly after our second interview, Jean experienced an episode of cardiac arrhythmia while having lunch with a friend. She lost consciousness and was rushed to the local emergency room, where she was hospitalized. She underwent surgery for a pacemaker implant and recovered well from the procedure. Her eldest son visited during the hospitalization and was attentive to the array of decisions that needed to be made. Jean decided she no longer wanted to live alone. She moved to a studio apartment in a retirement facility where all her meals were provided for her; the structured activities and entertainment on the grounds assuaged some of her isolation.

But in spite of the support from a few loyal friends, Jean experienced mounting anxiety and depression. She was again hospitalized, this time for psychiatric care. Upon her discharge, she transferred to a smaller board and care home.

Although Jean played an active role in the decision making, her adjustment to the new residence has been mixed. She felt terribly alone in both her condo and studio apartment, and the presence of others in this more intimate setting has alleviated some of her anxiety. She still attends the weekly Alzheimer's support group and received a great deal of encouragement from group members in her decision to give up independent living. Yet a feeling of loneliness is pervasive for Jean. Sometimes one can find solace in the presence of others and, to some extent, not living alone has helped her. However, Jean finds it increasingly difficult to track conversations while in groups of people and, because of the memory loss, sometimes cannot remember recent interactions or visits with friends. Thus, she does not always experience the sustained comfort derived from enjoyable time spent with caring people.

To the extent that I am able, I feel the aloneness of Jean's world. What becomes paramount in the lives of many people with Alzheimer's disease is constancy in relationship. There is an intensified need to rely on someone who can help to bring structure and safety to the world. There is a dependence and a need for security that are tempered by a desire for independence—a consciousness that says, "I will be my own person." But who am I becoming? What do I understand about my own person?

I reflect on the times when I have found comfort in my own company—in a relationship to self that, in the face of uncertainties or the absence of others, can be sustaining. Although my own psychological blind spots have been deeply unsettling at times, I generally experience security in a feeling of knowing myself—of having been with me for my entire life. In this way, there is some constancy in the companionship of *a comfortable sense of identity. Yet for Jean, it is this very sense of self that is undergoing such transformation. For many people with Alzheimer's disease, their confrontation with this changing identity is buffered by the familiarity of a spouse or family who mirrors back to them something of their own self. They are so well known by that spouse or family system that the system can sustain their sense of identity. Embodied in another person is the intimate, collective history of a lifetime that somehow remains while the very fabric of the individual with Alzheimer's disease is being woven into unrecognizable and unstructured patterns—still unfolding, but the pattern is no longer clear; the weave has unfamiliar variations. Rather than a well-worn and comforting shirt, the identity has patches of a somewhat different or alien texture.*

In those times, it is kindness that we hope for most. We hope that someone will be gentle with us as we face uncertainty—someone we can trust. It becomes essential for someone living alone to develop a supportive community, and Jean has this. Her friends from her various support groups, her psychologist, her long-distance family, the staff at her facility, and I all care about her a great deal. Yet, since the death of her husband, who is the anchor? Is anything or anyone certain? As any of us would, Jean feels this loss of security deeply.

Jean has many strengths—ones not commonly given tribute in our society. She has acknowledged her need to be dependent on others. In the face of vulnerability, she has opened a door into her life that has allowed people who were once strangers to become her friends and care providers. My own relationship with Jean has a fluid quality: she has been my mentor, my peer, my friend, my patient. This fluidity within the relationship defies traditional boundaries and role definitions. Yet this is the lesson. Jean knows I am significant to her because she feels positively toward me. As her disease progresses, I am less and less defined by any one role or identity and more defined by the cluster of thoughts, feelings, and expe-

riences that she associates with me. Likewise, as I continue to know Jean, I too am aware that she is important to me. But due to the effects of Alzheimer's on her, she is less a specific fixed identity in my mind and more a valuable and ever-changing person who engages me in an evolving series of experiences, feelings, and insights.

Jean teaches us a powerful process. Her willingness to look directly at herself while the very image reflected back to her undergoes transformation is testimony to her continued determination to remain connected to an experience of self. Yet, as with any crisis—whether material, physical, or psychological—if our self-concept is overly rigid, we will remain stuck in a resistant posture, unable to engage completely in life as we knew it and unable to learn our bearings and move forward into a new landscape. Jean has spent many years of her life in self-reflection, trying to understand her inner world. Although this process is very painful and frightening for her at times, it continues. As time goes on, I imagine Jean still in the audience while simultaneously on stage—watching her own life from an ever-shortening vantage point, while continuing to act in it. In this vision, she moves toward a point where she grows more and more into one perspective, the observer and the doer gradually merged into a singular awareness. And in this process, may her flexibility, and the kindness and goodwill of others, prevail.

Bob

"Losing my driver's license was like somebody cutting off my arm. I lost something that was a part of myself."

I met Bob at the first session of our Alzheimer's support group. He was referred by staff members at the Alzheimer's Association's Morning Out Club who noticed his despondency and hoped that the group could provide a place for him to air his feelings and frustrations. Bob was a retired mechanical engineer and an accomplished woodworker. He took great pride in his inventiveness and in his ability to outsmart challenges. Alzheimer's disease was terribly demoralizing to his sense of competency, and he was depressed. Social support is a well-known prescription for depression, and—prompted by his wife, Erika—he was practical enough to accept a trial. But it was a bitter pill to swallow.

Although Bob attended the support group regularly, he was reluctant at first to disclose much about his condition. His considered questions about the disease and his burdened countenance made it evident, however, that a great deal of commentary ran through his mind. The attention to detail so critical to Bob's profession as an engineer also permeated his *manner of speech; each piece of information led to the next in an elaborate construction of story that seemed both mechanical and methodical. Yet Bob had a depth of emotion that, when accessed, could well up into tears. He was grieving enormous losses.*

Bob persevered with the support group, but his ambivalence persisted. "I feel like I'm as friendly there as I ever get," he acknowledged, "but most of my life I've been a loner."

Bob was raised in the Midwest and recalls reading books in solitude when other children played in groups. Although friendly, he enjoyed a sense of self-reliance. His profession later served to provide a social structure, but his work acquaintanceships were based on collaborative ingenuity rather than interpersonal connections. At 70, after being diagnosed with Alzheimer's disease, Bob was suddenly and reluctantly catapulted into new social relationships. Once a month, the support group members and their spouses socialized and went to museums, concerts, potlucks, or parks together. Although Bob participated in these outings, his morose wit conveyed his reluctance at being included: "I enjoy all the people in the group," he remarked to Erika, "I just wish I'd never met them."

I recognized the struggle between despair and perseverance underlying Bob's predicament. He valued the group, but he wished he weren't a member. He didn't want to be linked to anything that was connected with Alzheimer's disease. I sensed that despite our positive rapport, there were times when he felt this way about me. Over the previous two years, our relationship had weathered significant challenges as we confronted the losses imposed on his life. But we communicated well with each other, and trust grew out of our candid dialogue.

Bob had a dry and mischievous sense of humor. As the two of us sat in his bright, sunlit kitchen, I reminded him of

*my reasons for interviewing people with Alzheimer's disease.
"You've come to the wrong place," he responded bluntly. "I
don't have it." Bob's deadpan expression surrendered to an
eruption of chuckles and his face flushed a deep red. As much
as I enjoyed his swift retorts and playful bantering, in our
moment of laughter, we probably both wished I had knocked
on a different door that day.*

The Onset

At the beginning, I didn't realize what was happening. Erika
realized that something was wrong. I wasn't aware. It's not
something that comes up and hits you in the head right away.
It's something that comes up behind your back. It's so insidious. It was three or four years before I realized that there was
a problem. I don't think I had ever really heard of Alzheimer's.
I had no concept of what it was. I didn't know that anyone
could have anything of that nature, and I just thought it couldn't
happen to me. But I started to go more inward. I didn't want to
go out.

At the time of onset, I would forget things. Erika would ask
me to go to the store and get something. I'd forget to buy it and
I'd get something else. I would do things that didn't fall in line
with what I had been doing before. It was a complete change
from how I had been previously. When I was working as an
engineer, all it took was one or two little triggers and I could
visualize and design all sorts of things. I helped develop the
pulse rockets that were used for the lunar lander. That's one of
the things I'm proud of, but I couldn't do anything like that
now. I also did some work at Oak Ridge *(where the atomic
bomb was developed)* and I'm sure that didn't do me any good.
God knows what kinds of chemicals I was exposed to. I can't
think of any other reason I would get this Alzheimer's.

An insidious onset is one hallmark of Alzheimer's disease. Scientists believe that the disease process in the brain may begin long before outward symptoms appear. But Bob's initial memory problem was also accompanied by a change in mood: his previous level-headed disposition began to give way to withdrawal and despondency. Major depression can also cause memory and concentration impairment and must be ruled out or treated before attributing symptoms to Alzheimer's disease. Bob was initially prescribed a trial of antidepressants in the hope that this would remedy both his mood and his cognitive losses. But when both conditions were unresponsive to treatment and his memory continued to worsen, the diagnosis of Alzheimer's was more certain.

Symptoms of depression include pervasive feelings of sadness, hopelessness, and helplessness; thoughts of suicide; a sense of guilt or deserving to be punished; changes in sleeping and eating patterns; and a pervasive disinterest in previously enjoyed activities. It is unusual for people with Alzheimer's disease to have major depression characterized by a cluster of these symptoms over a prolonged period of time. However, as Bob experienced, it is quite common for people to have one or more isolated symptoms of depression intermittently over the course of the disease. Episodes of discouragement are an understandable response to serious illness and can fluctuate as the disease progresses. Antidepressants may ameliorate these symptoms and alleviate some distress, thus enabling people to cope more effectively with the disease.

But despondent feelings do not always respond to medicine, and antidepressants cannot cure Alzheimer's disease. As Bob's Alzheimer's symptoms became more pronounced, he continued to experience difficult spells of discouraged and hopeless feelings. Accustomed to being a protective and capable partner, he was concerned about the effects of the disease on his marital relationship.

Shouldering a Burden

Poor Erika has to do all of the driving, thinking, and putting things together. Instead of my doing things for her, now she's the one who jumps in and gets things done. I can't do the things to help her that I could do previously. I wish that I could do more so that the burden wouldn't be on her shoulders. She has to be burdened with all of the things she does. I used to do the taxes. She never touched them, but now she takes care of them. Since Alzheimer's, there are so many things that I can't do. I try to help when she wants it, but I don't initiate very many things. I'll think of what to do, but then I'll fiddle diddle around and forget about it, and then start again and then forget about it. Erika tells me I should really get out there and do this or that. And I do. I mop the floor when she wants it done. Or I clean the patio. All I can do is to have her tell me what kind of help she needs.

Sometimes I give Erika a hard time just to be nasty. I guess it's because I'd like to be doing things myself instead of having someone telling me to do this or do that. I'm a little boy now. I have a mommy to take care of me. It's not a very good feeling. I'd much rather be out there doing something else. I don't know how it is with others, but that's the way it is with me.

Unless we've talked about something for a while, I will come back and ask Erika the same question over and over. I don't do it deliberately. It's just that each time it's like a new idea. And then I realize that it's not new and maybe we've talked about this more then once. She's patient. She doesn't say, "Bob, you know you've already asked that." She just gives me the answer again. Maybe after three or four times, I remember.

My tightest friendship is with Erika. I think we're growing more tightly together. I give her a hard time, but I'm just teasing. She ignores it. I guess I'm letting off some steam and frustration. She's awfully good to put up with me. I wouldn't ever think of doing anything that would be really detrimental to her. I'm lucky to have her. There aren't too many Erikas.

Bob felt very conflicted about the changing roles in his relationship with Erika. While he deeply appreciated her, his resentment toward his disease occasionally spilled out into episodes of stubborn rebellion in response to her suggestions or requests. He did not have an outward temper; rather, he was prone to brooding. Erika's life was primarily focused around Bob's needs, but Bob sometimes felt like his own life was surrendered to her: she did all the driving, made most of the decisions about their activities, and took care of the household business. Although she attempted to include Bob in much of the decision making, he felt there was little left for him to control except his occasional protest. He knew Erika was very capable, but he did not want her to have the burden of responsibility for his well-being, nor did he want to renounce his long-standing self-reliance. Yet, despite this ambivalence, he courageously acknowledged his limitations and was remarkably frank in recognizing his disability.

Acknowledging the Disease

During our conversation, the phone rang and Bob answered. The caller was selling a new phone company.

"I don't use the phone much." *Bob tactfully tried to work his way out of the discussion. The caller resumed the sales pitch, and this time, Bob tried a different angle.*

"I have Alzheimer's disease and I can't keep track of anything. My wife takes care of all this." *Still undaunted, the caller continued.*

"No, she's not here," *Bob replied. Another inquiry followed.* "She'll be back sometime this afternoon."

When Bob returned to the table, he was nonchalant. Rather than being annoyed, his foremost thought was to try to remember the message long enough to give it to Erika upon her return. He was matter-of-fact about disclosing his diagnosis to the caller, as he felt he needed an explanation for his disability.

Why shouldn't I tell people about Alzheimer's? Why hide it? Otherwise they'll think, "What is he doing?" I didn't tell people in the beginning. It wasn't until the Alzheimer's became more prevalent—when I couldn't do things that I had done previously. The effect of the disease is basically that I really can't do very much. I do things more slowly now, and there are certain things that I had better be smart enough not to try at all because it might be dangerous. Whatever I start to do, I botch it up unless somebody is watching me all the time. If it's something that I've done previously, I can get through. But if it's something brand new, it takes me a while. There have been a couple of things that I would like to do in the woodshop, but I just keep putting it off and putting it off. I just don't have the additional zip to go in and do things, and I don't want to cut any fingers off and send them flying across the room!

Other times I feel like I do things just great. Like when I go on walks. But of course, I fell one time. I don't know why I didn't put my hands down first. Instead, I hit my head right on the curb. I haven't been walking very much recently. I'm safer in the house. I'm not going to make any mistakes, and I don't have to go out on Erika's arm. I spend a lot of time reading at home. I can still comprehend what I'm reading.

Although Bob was candid in discussing his diminishing abilities, it was evident that the whole set of circumstances felt very demeaning to him. Alzheimer's disease was an assault to his long-standing pride in his abilities. Rather than confront his inability to accomplish tasks in a previously skilled manner or chance having another accident, Bob with-drew and limited his activity. Although he acknowledged Erika's tenacity in seeking activities for him, he also felt a great sense of indignity in the loss of his independence and autonomy. Nowhere was this more pronounced than in the revocation of his driving privileges.

The Loss of Freedom

The loss of my car is a big change. For some reason, I still feel like I should be able to drive. I went in to get the written test and the young lady apparently passed me. Then I was called in, and a guy drove around in the car with me. I failed the test. There was no discussion of what I missed or did or didn't do. I'd like to see that happen to him.

I've wondered whether I could appeal to another part of California and get a license. I feel that I could still drive as well as anybody. I'm scared when Erika is driving. I would rather be behind the wheel. Erika's a good driver, but not as good as me. I don't drink and I've never had an accident. I've never scraped the fenders or put a scratch on the car.

We went a thousand miles on a trip recently, and Erika did all the driving. Previously, we would drive an hour or two at a time, pull over, change positions, and away we'd go. So having to do all of the driving put a big strain on her. It made me strained, too, because I was keeping my eyes open every minute and watching everything she'd do. I still have a good sense of direction. I think it's from when I was in the Boy Scouts; I learned north, south, east, and west.

The complex issue of driving often creates considerable distress for Alzheimer's patients and their families. Many equate driving with independence, freedom, and basic compe-tency. But research indicates that the symptoms of memory

*loss, disorientation, and changes in spatial perception com-
mon to Alzheimer's disease may result in drivers becoming
lost, misjudging distances, forgetting basic rules of the road,
becoming more easily frustrated, or having slowed and incor-
rect reactions when making the multiple quick decisions
needed to drive safely.*

Yet despite these facts, it is very difficult for some people
with Alzheimer's to recognize the impact of the disease on
their driving performance. Because Alzheimer's affects people
differently, it may be relatively safe for some people to drive
during the early stages of the disease, whereas others have a
pattern of impairments that could place them or others in
danger. It is essential that anyone with a diagnosis of Alzheimer's be tested by the department of motor vehicles to
evaluate the effects of the disease on driving performance. In
some states, such testing is mandatory. In others, testing is
entirely voluntary. Because the disease is progressive, these
evaluations should be repeated every six months. But with no
consistent guidelines, many families find the issue of driving
to be an exhausting and heated debate.

Once diagnosed, some people eventually forfeit their driving privileges voluntarily. A doctor's recommendation, a family member's urging, an accident, or the diagnosed person's
own fear of inflicting harm are convincing enough motiva-
tions in this often painful decision. Others, however, lock a
firm grip onto the steering wheel and refuse to let go until
forced to by law, family intervention, or injury. As Bob spoke
of his severed privilege, his memories illuminated the deeply
rooted origins of his attachment to driving.

I was about eight or nine years old when I first drove. My
buddy down the alley had a father with an Oberlin. It had resin
glass and a circular thermometer on the hood. It had two seats
in the front and two in the back. That's what they used to go to
church in on Sundays. There was a service station about a

block and a half away from the alley where the car was kept. I'd save up enough money—a nickel or dime—and I'd get some gas to drive the car down the alley and then back into his garage. One time I saved up enough money to get a gallon of gasoline. We lived near the edge of town, and I drove the car down the alley and to the end of the street. I made a left turn and made a right turn and went out into the country and then back. I taught myself even when I was a little kid driving down the alley and back that if you have an accident, you're not going to get to drive again, so you better be careful. Our parents didn't know anything about this! I felt great. It was such freedom!

I'd like to be on my own. My independence is being taken away from me. Losing my driver's license was like somebody cutting off my arm. I lost something that was a part of myself. I lost my freedom primarily. Driving gave me the freedom that I am in control. If it comes to you, you'll realize what it's like to be deprived of your freedom by having to wait for someone to take you from here to there.

Indeed, as Bob spoke, I recalled when a broken leg had separated me from the driver's seat. I felt vulnerable and excluded from the dominant pace of the urban society around me. Dependent on others for transportation, I also was robbed of a degree of privacy and autonomy—someone always knew where I was going because someone always took me there. Not only did the loss of a license make Bob even more reliant on Erika, it also deprived him of that satisfaction so central to his childhood memory: the exuberant and sometimes mischievous expression of free will that can shape so many dimensions of identity.

Bob felt emasculated and punished by his loss. But this experience of demoralization is not gender specific. Both men and women can struggle with enforced dependency, with the fear of being a burden, and with a sense of diminished choices about fundamental matters in one's own life. Through his experience in the support group and Erika's counsel, Bob gained some consolation that he was not singled out for this disease. Yet, with such blows to his self-concept, it was hard not to take Alzheimer's disease personally.

I think we all realize what is happening to us. We really don't curse and swear, "Why me, why me?" In general, all the people we get together with now know what it's like. Unless we sit down and try to discover what it is that's different, you'd think nothing's wrong with me. Some people with Alzheimer's disease do things that are kind of wild, but most of us don't.

I used to think, "Why me? Why did it happen to me? What did I do? Did I eat something? Did I stay out too late?" But does anyone really know why I have it? I haven't been asking these questions recently. I've accepted it now. I don't like it, but I've accepted it. My wife has helped me accept it. She'll talk with me, and she's there for me.

You cannot physically fight Alzheimer's. You have to acknowledge what's going on and get with other people who have the same problem. Help each other as much as you can. I hope that people working in this field will be patient and learn to put the pieces all together. We need to keep hoping that someone will come up with a cure for this, but I doubt that anyone will come up with one in my lifetime. It would be great to go back to being a human being again. And it would be wonderful to get my car back!

After our interview and the delicious lunch graciously supplied by Erika, Bob suggested we take a short drive around his neighborhood so he could show me his walking routes and some of the surrounding scenery. I was pleased that he initiated an outing. Although I would do the driving, it was clear that Bob would be in charge of the navigation, and I noticed the seriousness with which he assumed command. I also became acutely aware of the somewhat decrepit condition of my car, which was overdue for a tune-up. Bob's attentive presence compounded every sputter and lurch the vehicle made, and my mind began a series of rapid associations with teenage driver's training classes.

"Turn left," Bob intoned. "Go right at the stop sign and then up the hill."

Half expecting him to produce a score card for my driving performance, I laughed to myself in appreciation of the equilibrium established by this pleasant tour of his neighborhood. Although I had directed the routes for our two-hour conversation, in these few moments of touring, Bob had effectively repositioned himself behind the wheel.

A year later, I revisited Bob and Erika in their home. The interior seemed even brighter than I remembered. I am sensitive to environments, to the placement of objects, and to the presence of color. I noted that Erika had moved a few pictures around; pleased by my attentiveness, she pointed out the new living room upholstery. I appreciated her quest for the solace and inspiration derived from creative endeavors. Bob proudly pointed to a high ledge that displayed three fish prints she had created in a class. Produced by carefully painting a fish carcass with colored ink and then pressing it onto special rice paper, the end result was strikingly beautiful. It was extraordinary that she could render a dead fish lovely.

Although I had not witnessed a dramatic change in Bob's disposition over the preceding months, I had observed him becoming sleepier in the support group and, in general, more withdrawn. He was starting a new antidepressant, and Erika thought it was helping him stay a bit more alert. As in our first visit, I reminded Bob of the value of his contributions *and the purpose of the eventual book. This time, there was no wry denial of his disease. Rather, his response was emphatic: "A lot of people have no idea what this disease is all about— no idea what we go through and the things that we feel we're losing." Recognizing the momentum behind his comment, I asked him to elaborate.*

My language has slowed down. I don't get the message across as well as I used to be able to. It's like someone is always holding me back. I have to stop and think and hope that I get the right words out. But the biggest thing of being held back is not being able to do the things that I used to do. Previously I was able to do almost anything I wanted to do. Now, I'm limited. All kinds of things have been taken away from me. That's the biggest frustration.

The theme of lost abilities is familiar to many people with Alzheimer's disease. To some extent, it is common to the experience of other disabilities as well. But the progressive nature of Alzheimer's exacerbates this challenge. Many people with Alzheimer's express a wish to halt the disease at one (mild to moderate) stage, thereby affording a time of adjustment to, or compensation for, a defined disability. Life could then proceed with some hope of stability. Although it is remarkable how successful many patients and families are in addressing each challenge along the continuum, the erratic nature of Alzheimer's disease does not often allow for a significant period of consistency. More commonly, people are asked to master the perpetual presence of unpredictability in their lives. As Bob reiterated his own frustration with these circumstances, he again expressed his concern and deep appreciation for Erika.

Erika is so good and she takes a lot of knocks—not physical, but knowing what is going on with me, and knowing that she can't really do anything but temper it somewhat. I know it would be hard for me if this were happening to her. We've been married 50-some years now.

I'm hopefully still tender with Erika. I would like to be able to help take care of her. She has always been a brick. She is a brilliant lady. Thank goodness she is level-headed—no odd ups and downs. I know that what she does is going to be right and is going to be of benefit to me and to her.

As Bill was impressed with Kathleen's rocklike stability, Bob appreciated Erika for her solid strength and fortitude. But as care providers will attest, sometimes fortitude erodes to expose frayed and fragile conditions.

Although Bob could not easily obtain respite from his own condition, Erika wisely recognized that someone had to maintain the energy reservoir. She needed time out and arranged a weekend away at a local inn for solitude and regeneration. Tom, their only child, flew in from Colorado to stay with his father. In the support group, Bob expressed concern about the plan. He was worried about Erika being out alone in the city, especially at night. He was reluctant to discuss any feelings of personal insecurity at being separated from his wife, especially given the company of his son. But as the weekend unfolded, an alarming turn of events confronted Bob with the unforeseeable nature of his own vulnerability.

We picked Tom up at the airport and then dropped Erika off at the inn. On the way home, Tom was driving. But I really wasn't sure who this guy was who was driving me. It took me some time. We had been talking, but it wasn't making much sense to me. I think I had been looking in one direction most of the time, watching the road. But as I turned to look at him, I suddenly realized by the profile that it was Tom. I realized that I had been riding with my son, and he had been doing a damn good job. It was such a shock to me because for a while, I didn't even know who he was. I think I told him after we got home that I really hadn't known he was Tom. I wanted to wait till we got in the house because I didn't want to upset him while he was driving. I didn't know what his reaction would be. He was pretty calm. He didn't make a big thing out of it. I don't know whether this occurs to other people with Alzheimer's or not.

The next morning when he and I were walking around, it was beginning to come together. Maybe the whole incident was good because it reinforced that I have a son—not something nebulous, but he's right there. When I was working in aerospace, I was in the office a lot, and there were times when we were separated. Hopefully, I'll get a chance to see him more often now. That would be nice.

I reassured Bob that indeed, the inability to recognize a significant and deeply familiar face probably happens more often to people with Alzheimer's than we realize. He shook his head knowingly, "And they don't want to say it."

"You might be right about that," *I concurred.*

I then reminded him of a recent episode in the support group when Bill courageously shared a startling episode of not recognizing his own face in the mirror.

"I'm too ugly to forget my own face." *Bob cracked a smile at his dry retort. (No one would describe Bob as an ugly man.)*

I felt both respect and gratitude toward Bill and Bob for sharing their disturbing experiences with the group. We often hear from both those with Alzheimer's and their families that the sign of real devastation will be when they can no longer recognize their own loved ones, let alone their own selves. Yet this symptom, occurring while Bob and Bill still had significant remaining abilities, diluted the dread associated with this Alzheimer's trauma. Although shaken, both the men and the other group members recognized that they could survive one of their worst fears.

The inability to recognize faces is well known to neuroscientists as "prosopagnosia." Thought to be rare in the earlier stages of Alzheimer's, the condition is more prevalent as the disease progresses. For Bob, with enough time and a few cues (conversation, different views of his son's face), the pieces fell together to form a recognized pattern. But until that point, Bob was being chauffeured by a stranger. He denied being frightened, having trusted that Erika placed him in good hands. Rather, the trauma existed at the moment in which he recognized his son. His disease had temporarily disguised a family member as a stranger.

Although Bob was remarkably calm during the whole process, other people with Alzheimer's find this situation terribly frightening and may furiously insist that their spouse is an impostor or their long-standing neighbor an intruder. Or, unlike the condition of pure prosopagnosia, some people may recognize a face as familiar but confuse that face with someone else's, often someone from their distant past. Hence, a husband becomes a father, or a daughter becomes a long-lost sibling. Given these disturbing experiences, it demands an extraordinary amount of trust for people with Alzheimer's to surrender their welfare to an unfamiliar face. And a profound patience and resiliency is required from those dedicated to providing their care. I asked Bob how he felt he was coping with these demands.

"I'm frustrated," he said. "Alzheimer's is not something where you can turn a switch and say, 'Aha. I fixed that.' But I just live with it, and hopefully there will be a cure. That would be the best thing."

Although Bob's engineering had helped propel men to the Moon, the mechanics of Alzheimer's disease were daunting. Bob had long been accustomed to applying his ingenuity to effect successful and satisfying outcomes. The boyish sense of mischief, adventure, and mastery so vivid in his childhood memory of driving had found expression in the adult pursuit of more sophisticated exploration and invention. But sometimes it is the earlier thrills (or traumas) of our life history that evoke our deepest feelings. Although Bob has shown extraordinary rationality in relinquishing control of many valued responsibilities and recreations, he remains most bitter and demoralized about the loss of his driving privileges. If he could take anything back from Alzheimer's disease, it would be the steering wheel of his car and the feelings of freedom, autonomy, and personal destiny it afforded him.

It is a particular challenge to be a self-described loner with Alzheimer's disease. Although some previously reticent people become more gregarious and uninhibited with the disease, other more solitary and independent individuals find the enforced social exposure of repeated medical examinations, structured Alzheimer's groups, specially designed activities, and an ever-present caretaker quite intrusive and insulting. When a treasured sense of privacy confronts a growing need for protection, the result can be a considerable amount of aggravating friction for all involved.

Given his disposition, it is commendable that Bob accommodates the enforced society of Alzheimer's disease. Sometimes, professionals and families automatically prescribe the prevailing treatment for Alzheimer's—social support and stimulation; increased supervision; safety precautions (such as the termination of driving)—without paying attention to the side effects of this prescription on each unique patient. Clearly, not everyone will comply with these recommendations; in the best of circumstances, we introduce one intervention at a time and allow for a period of adjustment before recommending another one. Bob is stoic and cooperates begrudgingly with most of the recommendations. He continues to participate in the Morning Out Club, in the support group, and in various outings that Erika organizes. He does find enjoyment in life. But he is still prone to despondency and admits to episodes of a "poor me" *attitude. Antidepressants have not remedied this mood. They do, however, assuage the troubling problem of drowsiness and help him to be a bit more alert and engaged with the world.*

More than three years have passed since I first met Bob in the support group. Alzheimer's disease has taken away the autonomy that he used to enjoy and confronted him with the complexities of human relationship and interdependency. I have been moved by this relatively reclusive man's expressed appreciation of Erika and his genuine concern for the welfare of his fellow group members. Although he may seem indifferent at times, it is clear that he cares deeply. I don't doubt, though, that if given a choice, Bob would reach for the solace of his car over the society of his friends. When one's freedom is lost, it's not easily replaced.

Booker

"I'm at ease, and I'm one of those people who is blessed in that way."

In a quiet cul-de-sac, surrounded by dry southern California hills, rose bushes bloomed in a carefully tended garden bed under the front window of Booker's residence. After the sudden death of his wife in 1993, Booker moved from New Jersey to live with his daughter, Brenda, and her husband in their San Diego home. It was then that Brenda noticed changes in her father's demeanor and overall abilities. Always a man of gracious disposition, Booker became forgetful, frustrated, and irritable. Although some adjustment strains would be understandable given the loss of his wife and the dramatic change in his living situation, Booker's condition did not improve with the passage of time. Brenda, confused and fatigued by her father's increasingly irrational tendencies, insisted that he undergo a comprehensive evaluation for memory loss at our UCSD Alzheimer's Disease Diagnostic and Treatment Center. At age 82, he was diagnosed with Alzheimer's disease.

Booker was very composed throughout his evaluation. When asked about it later, he had no specific recollection of *the process. But he understood the outcome and seemed to accept his diagnosis with a reserved dignity. Impressed by his poise, the diagnosing physician referred Booker to me for an interview. After I had an introductory phone conversation with Brenda, she obtained her father's consent, and we scheduled a home visit. With the source of his problems clarified and the prescription of an antidepressant to help relieve his anxiety and irritability, new household rhythms were gradually being established, and life in the home was starting to calm down.*

Brenda greeted me at the front door. While Booker finished his breakfast, I sat quietly in the immaculate living room organizing material for our interview. The home reflected an attention to detail—a certain appreciation for order that I was self-conscious about disrupting with the clutter of my accouterments. But I did not have long to worry before Booker emerged from the kitchen. As Brenda introduced us, I immediately noticed her father's elegant stance. Booker's tall, lean stature conveyed a demeanor of self-respect. As we reviewed the purpose of the interview, he considered the information attentively, responding with formal but welcoming propriety: "I'll be pleased to cooperate with you."

Booker's subtle bow and invitation to dialogue established an elegant ritual for our conversation—an understanding that he would give me his attention and I would reciprocate with mine. I reflected on the harried society in which I live: How often do two people take the time to engage in extended, uninterrupted dialogue, to truly meet each other and experience the privilege of this uniquely human form of interaction?

Born in 1914, Booker was raised in rural Virginia. He is quick to recall that his own father was emancipated from slavery at age nine and grew up to become a Baptist minister and a major influence in his life. During World War II, Booker worked in Virginia in the shipbuilding industry doing steel construction on cruisers and aircraft carriers. He later moved to a suburb of New Jersey, where he worked in a foundry that made printing presses. There, Booker and his wife raised Brenda, their only child. Following in his father's footsteps, Booker became active- ly involved as a deacon of the Baptist church.

As Booker's life story unfolded, the cadence of his speech lulled me into a pleasant time warp. There was no rush in this exchange: the pace evoked images of a white wrought iron park bench situated under a grove of shade trees on a slow, humid, Southern day. There were stories and lessons to tell—not long- winded monologues, but short tales with a moral to teach. Booker's message was part biographical, part philosophical, and part sermon. His satisfied laugh periodically refreshed us like a long, cool drink of lemonade, and as I adjusted to his rhythm, I appreciated the privilege of meeting a man whose life path probably never would have converged with mine were it not for this crossroads at Alzheimer's disease.

Shortly after his move west, Booker and his daughter visit- ed Jerusalem, where they were baptized in the river Jordan. This journey was foremost in his mind as we began our conversa- tion. As the scrapbooks came out, I commented on the similar- ity between San Diego's and Jerusalem's arid landscapes. Booker smiled knowingly. I felt grateful to him for reminding me that people have many dimensions. How could I inquire about Alzheimer's without understanding what was meaning- ful to this man? The baptism was a pivotal experience for Booker. And as I listened to his reflections on living with Alzheimer's, it seemed that he had submerged personal control of his diagnosis and condition in the divine forces at work in his life.

A Mystery of Life

This disease is something that has happened to me and I accept it. Time brings this disease on and I've been going on a while now. You come to an age when your system changes. We're human beings. We're not divine. Time takes its toll on anything. A piece of wood will change over time. It's kind of a mystery, though. This illness is just something that has come into being that you can't really understand. How did it happen? But God is my foundation. My faith is solid. My faith is within me and it *is* me.

As Booker spoke, I felt a soothing sense of human integration with the natural world. Life had its seasons and cycles, its periods of generation, transformation, and degeneration. His illness was perplexing in its origins, yet he accepted that life contains a realm of complicated mysteries beyond human comprehension or reason. He had no recollection of any alarming incidents or peculiar changes indicating a disease process under way. Rather, at age 82, Booker accepted that like a full year's passing seasons, changes are expected when the circle of life is moving toward completion. His condition was a natural response to a well-used mind and body that just wore out with time.

Age does prevail as the single greatest risk factor for Alzheimer's, and some researchers argue that anyone who lives long enough will eventually develop the disease. But as our population's life expectancy increases, there are a growing number of elderly people passing into their tenth decade and beyond who may have small numbers of plaques and tangles in their brains, yet show no cognitive or behavioral evidence of dementia. Thus, although the disease seems natural in Booker's paradigm of the life cycle, it is not necessarily an inevitable component of our society's aging.

In a Daughter's Hands

As I heard the logical rebuttal to Booker's perspective in my mind, I recognized a state of grace inherent in his more naturalistic view. He did not welcome Alzheimer's disease, nor did he fear it. His philosophical surrender did not result in the loss of independence and control. Rather, it filled him with comfort and contentment. Nowhere was this more apparent than in discussion of his daughter.

I'm blessed to have a wonderful daughter. She's very good to me. She looks after me. I love her and she loves me. I can be a bit more at ease in my thinking because she takes care of things, and she knows how to do it. She looks out for my best interests. I sent her to school and to college, and now she knows how to take care of all my business. I depend upon her. I'm in her hands. I'm in my baby phase now, so to speak. So sometimes I call her "my mumma." Yes, she's my mumma now. *(Booker smiled appreciatively.)* I take my own bath, but sometimes she tells me what to put on that day, and I'm obedient. She sees that I'm dressed just right—makes sure my tie's on straight. Sometimes I lose things—maybe my pocket knife. But I have my own room. That's where I keep everything. My daughter takes care of my forgetfulness. Her husband is wonderful to me, too. People know my problem, and they look out for me.

There's nothing much that you can do if you have this disease. You have to let someone support you. My disease is not into it too far, but as it progresses, it hurts you more. When you're saturated with it, people have to lead you around and do things for you because you can't think or control your situation. I don't have to tell people literally about my diagnosis just yet. But I may have to instruct them sometime to know the procedures if something happens to me. But I've got my daughter now. She's my backbone. She's such a blessing to me.

Brenda confirmed that throughout Booker's marriage, while he had been responsible for the family income, her very capable mother had been in charge of all matters pertaining to the maintenance of the household and the personal care of its members. Booker was accustomed to this dynamic. Now, with the passing of his wife, his daughter became the "mumma" of the household, and he gratefully placed himself in her hands. This did not imply that he was incompetent or unworthy of dignity and respect. Rather, in the full cycle of life, it was the natural order that the younger generation return care to its elders.

But the natural order sometimes falls into chaos. Brenda was married in 1992. Nine months later, her father moved in with the newlyweds. Six months later, around the time of Booker's Alzheimer's diagnosis, Brenda's husband—a marriage and family counselor—required a kidney transplant. And the following year, Brenda returned to school to study business, while working full time. Pressed between the care needs of her recuperating husband and her dependent father, Brenda became a full-fledged member of what sociologists have termed "the sandwich generation." More commonly, middle-aged children are sandwiched between their growing children and their aging parents, but Brenda had enough people to care for, even with no children of her own. Fortunately, her husband, who recovered from surgery and is now retired, assists with Booker's care. "I'm hanging in there," Brenda reflected with a soft smile. "It's a little rough, but now it's better since we're dealing with one illness instead of two. When they diagnosed my dad, I thought someone had punched me in the stomach. But then I moved through the stages of acceptance very quickly because we needed to know what to do."

The Neighborhood House

An extraordinary amount of organization is required of adult children who juggle the demands of their own families, their professions, and one or both dependent parents. Brenda recognized that for the household to run smoothly and relationships to remain cooperative, a sense of routine needed to be established. She enrolled her father in a structured program at a local community center for seniors with cognitive and physical disabilities. Although Booker resisted at first, his daughter's respectful but firm persistence prevailed, and he adapted to the program successfully.

I don't have to get up early in the morning unless I'm going to the Neighborhood House. I put the alarm clock under my pillow and it wakes me up on time. Then I get ready to go gather with the other seniors in the program. I have to get up and shave and get myself presentable so I'm ready when the bus comes to pick me up. We have fun at the Neighborhood House. We have some classes and little programs there, and we have a chance to express ourselves. We go on walks and outings. It's a senior center, and the people are active. I miss it when I can't make it there. If I just keep active, maybe this illness won't progress so fast. If you're active, it keeps you going in your mind and you don't just stagnate. But time takes its toll on you if you hang around.

Although Booker accepted his disease, he was not complacent. By attending the Neighborhood House, he felt that he was doing something productive and therapeutic instead of passively succumbing to Alzheimer's. The senior center also provided an experience of social belonging that is so often diminished, or absent, in the life of a person with Alzheimer's disease. The past decade has seen a burgeoning of social centers with day-long programming specifically designed for cognitively impaired adults. These programs serve the dual purpose of creating stimulating and compassionate environments for participants and also providing families with much needed respite from the responsibilities of caring for their loved one.

Some people with Alzheimer's refuse to attend social centers. The childhood associations evoked by the structured activity, the discomfort of being grouped with others who are more significantly impaired, or the fear of separation from the lifeline of a familiar family member or spouse can all be significant deterrents. Booker's comfortable associations with a cyclical return to aspects of childhood allowed him to define the social program as a classroom where he was afforded an opportunity to use, and hopefully preserve, his skills. Rather than resisting the structure, he relaxed into the routine. He generally seemed relieved when someone else took charge.

A Touch of Love

As much as Booker gracefully relinquished some control over his life, he was also concerned that his courageous trust in others not be violated by insensitive or harmful care.

If I forget something, I want people to be mild with me. Do what you have to do, but you appreciate a touch of love rather than a touch of hostility. Hostility will cause you to rebel. Treat others as you would like to be treated. You wouldn't like me to be beating on you all the time.

Things come along sometimes that I don't like, but I just try to push them aside. Whatever comes up that I don't appreciate, I try to push aside. To a certain extent, you can't just treat me any way. If I have to come out of the corner, I'll come out of the corner. You just don't run over me. But with hostility, someone is going to lose. So, I'm not kicking up my heels these days or raising too much Cain. Sometimes I look at the youngsters out there and they think they know so much. They're hip. And they look at me and they think, "That ol' man don't know nothin'." That's what I think they have in their minds. But I've been down that road myself. I know they're going through a stage that I've been through.

I was brought up under the teachings of my dad. He was a pastor, and his instructions stayed with me. My daddy never yelled at me. He was a mild man, and he taught me how to behave. Those values that you have received, they are in you, and you can pass them on to others if they'll let you. I can't remember ever spanking my daughter. I just would talk to her. And that approach rubbed off on her. And now, she's mild with me. So you have to be careful how you train your children because they're watching you, and pretty soon they pick up on the way you handle things. If you sow bad, you reap bad. You can be hostile to a person and they can fear you, or you can be kind and loving to them and instructive to them. Then you will be planting some good seed.

As Booker tried to live peacefully with himself, he also hoped for continued harmonious relations with others. Over the course of our conversation, he often repeated the story of his childhood and his father's mild-mannered teachings and discipline. Booker's immediate memory loss rendered the story fresh to him with each recollection. And each time I listened, I heard more deeply the life lessons essential to his well-being now. As he returned to the vulnerability of his youth, he hoped he would once again be met with the parental sensitivity of his father. And, though Booker was indeed fortunate that his daughter and son-in-law carried on this tradition of loving care, other vulnerable seniors face much harsher circumstances.

Elder abuse is a widespread problem. Each year, cases of physical, emotional, and sexual abuse; financial exploitation; and neglect are being referred to authorities in ever-increasing numbers. Yet due to family secrecy, the incidence of elder abuse is grossly underreported. Although it is not clear how many of these elderly people have Alzheimer's disease, many studies report a higher level of abusive behavior when Alzheimer's enters into the family caregiving relationship. Contributing factors can be caregiver depression and stress, a diagnosed person's disruptive behavior, or long-standing relationship conflicts that become further exacerbated with the challenges of illness. Once revealed, abusive behavior may be remedied through increased caregiver education, emotional support, and respite or through medical evaluation and possible treatment of an Alzheimer's patient's persistently disruptive behavior. Sometimes, more severe circumstances warrant complete removal of the caregiver and the securing of a safe haven for the person needing care.

I recognized in Booker's household many of the elements that serve to safeguard against abuse: Booker's attendance at the Neighborhood House gave everyone some necessary time apart from one another; consistent and thorough medical care was available to evaluate and treat the disruptive symptoms of Alzheimer's; a long-standing positive relationship between Booker and his daughter was grounded in mutual respect and gentle care. Yet we should not underestimate the unpredictable twists and turns of Alzheimer's disease even in the best circumstances. It is not a question of whether stress exists but of whether we seek, and have available, the means to address it effectively.

A Sense of Satisfaction

Few changes occurred in the four months between my two visits with Booker. A sense of structure was established that generally met the needs of the family. The comforting predictability of routine seemed to override the upheaval and challenges of Alzheimer's. Also, although memory loss is the common denominator for all people with Alzheimer's disease, Booker's short-term memory loss is so profound that he may not suffer the painful self-consciousness that some patients experience. He can still be stubborn or irritable at times, but sometimes he forgets that he is forgetful. Perhaps this is a gift. As moments pass by and cycles run their course, he focuses on *what life offers, paces himself slowly, and dispels any apprehension with the certainty of his faith.*

I don't have any worries about the future. I'm satisfied in my heart, and I'm at peace with myself. I'm at ease, and I'm one of those people who is blessed in that way. I'm thankful for that. I'm a Christian and I try to live by the Book. I'm steeped in it now. You couldn't pull me away from it. If you have a solid foundation, you can be at ease. Accept life as it is. Accept the fact that you didn't come here to stay, and sooner or later you'll pass away. "Be ye also ready for ye know not when the son of man cometh to take you to eternity" *(Matthew 24:44)*. I'm ready.

If you can be at ease with yourself, you can be an example to others who you come in contact with. I try to treat people right. I'm not all excited over this disease. I'm thankful that I am as active as I am. I don't have to use a cane. My limbs are useful, so I'm blessed.

The doctors are doing the best they can for me. They gave me a few pills to take that are supposed to be good for the problem. You have to have researchers trying to understand this disease. To them I say, "Seek and ye shall find. Knock and the door shall open" *(Luke 11:9)*. You all are seekers, so keep seeking.

Although Booker understood the seriousness of Alzheimer's disease, he experienced little fear. The foundation of his beliefs and the structure of his family provided reassurance and routine. Although he still commanded respect, he accepted his need to rely on others and deeply appreciated their efforts to help him. There was a time to give and a time to receive.
I wondered if Booker had ever known anyone else with Alzheimer's disease, because our responses to various life circumstances are so often influenced by particular memories or

associations. When I posed the question, he paused thoughtfully and then told me a story.

In the rural Southern town of his youth, there was a woman who was quite aged. Her name was Katherine, but everyone called her Aunt Kitty. And she'd answer to that. Aunt Kitty walked with a cane, and she'd go walking with her cane out in the country down the long lane. But she was prone to forgetfulness and confusion and would sometimes get herself into predicaments.

One day, as Booker walked down the lane on his way to the mailbox, he spotted Aunt Kitty in trouble. There was a big ditch that ran the side of the lane—a deep ditch, as deep as Booker was tall. And when it rained, the ditch would become streamlike. People used to call it a canal. There had been some rain, and there was water flowing through the ditch, and there was Aunt Kitty stuck on a mound of dry ground, right in the middle of the canal. Now how she got there, Booker didn't know. But Aunt Kitty had a death hold on a bush there on that dry ground with the water running around her, and she wasn't letting go. Booker saw that he couldn't get her loose and carry her across himself, so he went for help. He rounded up two other fellows, and together they told her to let loose of the bush and hold onto them. They made a seat for her with their hands, and she clasped her arms around their necks. They brought her out to the lane—to dry ground—where she could walk. She was so grateful, she said that if she had $100, she'd give it all to Booker for getting her some help. So Booker felt blessed—blessed that he could help her when she was confused and in need, and blessed to receive her gratitude.

That's the story about Alzheimer's that Booker remembers to this day and says he'll never forget. As his disease progresses, he too may suddenly find himself alone, confused, or in a frightening predicament. If so, he trusts that just as he and his friends were there for Aunt Kitty, someone will be there for him to lift him across those rising waters to the security of safer ground.

Betty

"A person with Alzheimer's disease is many more things than just their diagnosis. Each person is a whole human being."

Before I ever met Betty, I was screened by her husband, Kurt. I had notified the San Diego community through the local chapter of the Alzheimer's Association of my interest in collecting verbal or written reflections from individuals diagnosed with Alzheimer's disease. Kurt saw the Alzheimer's Association notice and called to investigate my intentions. His wife, Betty, newly diagnosed with Alzheimer's, had not written any personal commentary on her condition. But my interest in her perspective intrigued him. Both he and Betty were retired social workers. The project of collecting narratives from people with Alzheimer's piqued his interest as a worthwhile humanistic endeavor.

I proposed that I interview Betty in their home as an alternative to her composing a written narrative. Kurt discussed the matter with Betty and, a few days later, called back with her consent. A delicate but purposeful review process was under way—much as one might investigate a potential new in-law. Both my physical presence in their home and my proposed *interview would bring me into their personal domain. With just caution, Kurt was opening a window. It wasn't until I met with Betty's approval that I would be let in the door.*

Kurt and Betty did not live far from my neighborhood. Perhaps, as social workers, we were drawn to the urban communities of San Diego where people lived amidst the complexities and creativity of city life. Kurt and Betty's neighborhood was originally established in the 1920s and 1930s to serve the community near San Diego State University, where they both had been on the faculty. In recent decades, the old family dwellings had been infiltrated by scattered apartment houses and condominiums. As I drove down their street and farther out to the hillside ridge, the intrusions were fewer, and hints of the historic Spanish-style architecture of San Diego shone through. When I arrived at Kurt and Betty's residence, their vintage stucco home—surrounded by cactus, yucca, and other plants native to southern California—spoke of history, of stability, of the passage of time gracefully encapsulated in architecture.

Kurt and Betty greeted me at their front door. Betty's closely cropped white hair; her large-frame glasses; and her deeply lined, tanned face gave her a bright, intellectual look. Although Kurt had been dominant and protective of Betty on the phone, from the moment I entered the home, it was Betty who took charge of the visit. This was a couple of long-standing partnership. Together, they emitted a sense of balance that conveyed a history of supporting, challenging, and compensating for each other. Kurt and Betty—the same height, age, and profession—were a team. I wondered how many thousands of people they had greeted at this door over the years. The ritual, though well known to both of them, was fresh, sincere, and inquisitive.

As I walked through the front door, I entered not only their home but an environment that had facilitated many professional and personal group meetings, many interviews, many opportunities to explore human connections to other people and the surrounding world. A feeling of welcome permeated the large, old ceiling beams and simple, comfortable furniture of their expansive living room. But we would not be meeting in the living room. Instead, Betty guided me to a serene guest room, where she closed the door and effectively shut Kurt out of this personal experience. She reclined into a chair, put her feet up on an ottoman, and relaxed into a mode of self-analysis. I sat opposite, and our relationship began.

As Betty spoke, I listened, as I had with Booker, to the cadence of her speech. She looked up and away—pulling pieces of text, memory, and speculation out of her mind—musing and postulating, assessing and objectifying. We were beginning a walk through the museum of her life. This would be a deliberate, considered pace, with time for comment and critique along the way.

The content of conversation is important, but if I don't accurately gauge the rhythm, I am less apt to be able to be an accompaniment in an individual's particular musical score. It's not hard to recognize when I'm off the beat. I feel it internally, as if I'm dancing the waltz to improvisational jazz. Although Betty's welcoming warmth conveyed a woman of deep feeling, this was not the rhythm of her initial speech. Rather, it was her analytical mode, herself as intellectual clinician, that guided us through our interview. It was as if she were becoming her own case study.

Betty grew up in a small industrial Pennsylvania town. Her alcoholic and abusive father contributed little, if anything, positive to the family, leaving the responsibility of livelihood to her mother and the children. Betty was determined to escape this burden of powerlessness and poverty. As a child, she worked in her parents' small-town family store. Through her interactions with the diverse customers, Betty learned to observe people—to become an astute evaluator of human behavior. These skills were revealed in her chosen profession and also served as the foundation of her own self-observation.

Early Signs

I turn 78 in November. I think it's just within the last year that I've noticed I have a memory problem. As a social worker, I dealt with people who had memory problems, so now it's a matter of being honest about my own.

I never was good at memorizing. I always knew that, and it didn't bother me. But recently, during lengthy evaluative testing, it was clear that there was something else going on. While taking the tests, I was aware that I just didn't have the usual ability to recall. I've needed to have that ability for most of my life, so I know when it's not there.

I've been in the same swimming class for two years, and I don't remember everyone's names. They just escape me. I have to listen to people calling other people by name so I can catch on. I find all kinds of ways of doing that! I've learned many ways to get around not knowing someone's name by relating to the person first and seducing them into doing more talking. If they talk about five minutes, pretty soon I have enough clues. I can usually identify more quickly with what people do, and where they come from, than their actual name.

There is always a startled feeling when I forget something that I could have talked about yesterday. I get upset with myself and think, "Oh damn it all." If I sit with it a while, maybe it will come back. An hour later the whole thought may return, but it's too late because the moment has passed. Sometimes I can say, "Oh I forgot to tell you this." But it's all a matter of timing.

Also, I know that I may take something away from home, put it in my pocket, and then be completely unable to determine what I did with it. I just don't recollect. Things like that happen sometimes. It's there. I'm aware that my memory doesn't work as well as it used to.

Betty's early symptoms of Alzheimer's disease could be common complaints from people of any age, especially seniors. Unlike Bill's more alarming impairments at age 50, when symptoms appear in later years, they are easily attributed to age. Family, the public, and medical professionals often dismiss mild signs of Alzheimer's disease as the "benign senescent forgetfulness" of growing older. Indeed, as the body's physical reflexes slow with aging, so do cognitive reflexes. The ability to recall known information or to learn new information may require more time and concentration. But despite effort, Betty could not retain material that she could have learned previously with practice and patience. The development of consistent episodes of memory loss that reflect a marked and ongoing change in previous ability is not an inevitable component of aging. When combined with difficulty in performing previously well-learned tasks, concern is warranted. Although individuals in all stages of Alzheimer's disease may experience day-to-day fluctuations in ability, the underlying memory problem does not disappear and cannot simply be attributed to a bad day.

Betty was pragmatic. She was not going to be duped by her own symptoms, and she intended to address them in a straightforward manner. But under her assured demeanor lay a component of vulnerability that was much more personal.

Family History

The other reason I'm especially sensitive to Alzheimer's is that my younger sister has it. It was about 10 years ago when she had the first hints of disease. When she was diagnosed, I thought that might be my eventual fate, but I hoped that it wouldn't be. She is presently in a small nursing home close to the valley where she grew up, near her friends. They're taking good care of her at this home. Her son tries to visit her and do all he can. The first time I went there, I thought, "Oh my gosh!" But the woman who runs it is a lovely person, and that helps a lot. So, my sister is doing as well as can be expected. I call her once or twice a week. She always recognizes my voice, but she has begun to lose the ability to identify old friends. If she sees them in person, it's easier. But it's obvious why that's so. There is more to remind you about a person when you can see them. If a friend has her hair done the same way all the time, you can say to yourself, "Well, I remember the hair."

I'm wondering how long this process will take before I have to be taken care of. That's the obvious concern. I know that some people progress in Alzheimer's very quickly and with others, it takes a long time. So I have two wishes: one is that it doesn't end up being too bad or that it's a very slow process.

Betty's family history of Alzheimer's disease made her more sensitive to the development of her own symptoms. It also placed her at an increased risk of acquiring the disease. When one has a parent or a sibling with Alzheimer's, one's own susceptibility is increased. With the exception of rare and specific genetic variations, however, family history by no means necessitates the development of Alzheimer's disease. Even among families, the rate of progression is not consistent, nor is the presentation of disease characteristics. The duration of life with Alzheimer's varies from 2 to 20 or more years, and behavioral and cognitive symptoms vary considerably among affected family members.

Teamwork

While Betty was hopeful that the course of her Alzheimer's would be slow and her symptoms mild, she was already aware of the support and security derived from her relationship with Kurt. Although the disease tried their patience, the marriage was a source of deep familiarity and comfort in the face of uncertainty.

It helps that I'm not alone in this. For the time being, Kurt is there to correct me on things, and I also correct him sometimes. But as long as Kurt can stay one step ahead of me, I'm not going to worry. We've been a team for a long time, so I guess I don't have much choice in the matter and neither does he. I'm more dependent now, and I've never been particularly dependent. I'm not glad about the fact that this is happening. But I know that I have no choice at this point. Sometimes when I forget something, Kurt has to get hold of himself and not get all uptight about it. Obviously, he's having to learn this over a period of time. It irritates me that he gets ticked off over something that I've forgotten, and every once in awhile I blow up. I'm sorry about it, but I just forget things.

I think it's sad that there isn't any history of teamwork in the marital or family relationships of a lot of people with Alzheimer's disease. I've worked with many different people and I know this is the case. There has to be a lot of trust. That's true of any good marriage. When I think of the marriages that have gone awry, they didn't have this element. By having enough trust, you can bear a lot of things; trust about making decisions that are really mutual, or trust that you can talk about behavior on one or the other's part. We have no problem pointing things out to each other.

The consistency and teamwork of their long-term relationship sustained both Kurt and Betty. It also, however, created challenges that are common to many families facing Alzheimer's disease. For better or for worse, we grow accustomed to the styles of interaction, personality characteristics, quirks, weaknesses, and strengths of another person. We know the choreography of a relationship—the ways in which we move gracefully together and the steps that can cause us to trip. In good teamwork, there are areas where one person may lead and the other follow. Yet while some relationships can adapt to changes in rhythm, others know only one or two different routines. Their less developed teamwork is thrown into crisis by any new percussion. Even the most flexible partners can tire of having to learn new moves. Families may struggle to maintain the same old dance while hoping that the exasperating beat of memory loss will just fade away. Hence, as Betty described, while you do your best to keep moving, patience can be trampled under tired feet.

I am at an advantage in establishing and building a relationship with someone after the onset of Alzheimer's. I'm familiar with the dynamics of the disease, and my task is to see how each family member is moving within its challenging rhythm. I expect that we may stumble while interacting with each other's styles. Sometimes we collide. I offer what I hope is helpful or believe is necessary to ward off crises. But over the years, each family has been an invaluable instructor in the many different ways of moving through this disease. I am always prepared to be both teacher and pupil in this kinetic process.

Letting Go of Responsibilities

Betty was well aware of the ways she and Kurt were having to alter their well-established routines. Both were still very active in community social work administration and consultation, but Betty knew that she no longer could take the lead in organizational matters. She recognized the need to pace herself differently.

I'm not taking on any new responsibilities. I find a way to get out of them. I do a few things now and then that I am still able to do. I've been involved in many social work organizations, and I still get about 10 professional journals a month that I skim. I might read an article, and it helps me feel in touch and involved, even if I'm not. I'm being very careful about getting too involved.

We keep our house open to guests and colleagues who visit from around the world. I haven't told any of them about my memory loss, and I think this is going to be my coming out time. If someone asks me, "Why can't you take on this project?" I'll say, "Well, I'm beginning to be concerned about my memory." I'll relate it to Alzheimer's disease. I'm not avoiding that. That would be an obvious reason for not staying connected in an ongoing, detailed fashion. I'll just simply offer to resign some of the responsibilities. People will be disappointed that I won't be able to do the work. But eventually, I'll need protection from myself—protection from not doing things that would be stressful. In the past, if something looked stressful, sometimes I'd say, "To hell with the consequences!" But I know that I can't say that anymore.

Although Betty was learning to moderate her activity, her statements revealed the challenge of staying connected to interests and purpose in life while also respecting new limitations imposed by Alzheimer's. Whether caused by the onset of disability, the birth of a child, an illness in the family, or a change of jobs, when the rhythms of our life shift, we have to pause, catch our breath, and adjust our routine. When necessary, relinquishing some responsibilities can be a relief. Yet the cause of the shift in Betty's life was a disease whose rhythm is complex and not always well accommodated in the predominant choreography of society. Although Betty knew the reasonable course to take in acknowledging her diagnosis and in disclosing it to others, feelings of uncertainty and caution permeated her more pragmatic demeanor.

Unpredictable Responses

In the past, people never uttered the word *Alzheimer's* for fear that they would catch it. They were defending against it. It used to be that way around cancer. But now Alzheimer's disease has a lot of attention, and the symptoms that are described scare people—that we'll walk blindly into a car because we're lost and wandering. It isn't necessarily true, but people get an idea.

When it comes to Alzheimer's you're not sure how people will respond to you. None of us like unexpected responses. People may brush you away because they are afraid of the disease. They may feel uncomfortable because they don't know what to say to you. It puts a burden on people to figure out how to respond. So, I'm not going to run around saying, "Hey, I have Alzheimer's. What do you have?" If you have a physical disability, it's quite obvious. But this kind of problem is more difficult to convey. The consequences are serious, but you don't want to have to spend hours talking about it in an impersonal discussion. That is true of any diagnosis of serious illness.

It's very different if you know that you are talking with someone who is familiar with the disease; there is a safety net when you talk with people who understand and care about your condition—people who don't step on your feelings or minimize your problem. When you forget something and somebody says, "Oh well, it's not important," maybe it was important. It shouldn't have been forgotten, but it was, and you need an explanation for yourself.

People may deny that they have Alzheimer's disease because they don't have the opportunity to talk with other people who are sympathetic and understanding and who will help them along in the whole process. That's a sad state of affairs. Anyone who has this diagnosis needs to have others with whom to talk.

Anytime we receive a piece of news, whether promising or discouraging, we go through a process of deciding with whom, and when, we will share the information. In the case of good news or relatively impersonal information, the process may be minimal. But when the topic is a diagnosis of Alzheimer's disease, it is not uncommon to hear those diagnosed and their families weigh the disclosure carefully.

Betty took the first step in disclosure by acknowledging the diagnosis to herself. Many people may initially, or more persistently, doubt their diagnosis. Family and friends can also go through periods of denial by ignoring the problems or invalidating expressed concerns as "problems that happen to everyone." Certainly, it would be difficult for some people to match their bleak images of advanced Alzheimer's disease with Betty's vitality and insight.

Yet it is this disparity between the general public's limited perception of Alzheimer's disease and the actual full spectrum of the condition that contributes to hesitancy in acceptance and disclosure. Many people with Alzheimer's refer to the sense of stigma associated with the disease when others assume they are incompetent, tragic figures or, as Betty so aptly described, bizarre and unpredictable in their behavior. Although the effects of the disease run from mild to severe, people who are newly diagnosed or only mildly to moderately impaired may fear the categorization and subsequent aversion that can come from our dominant conclusions about people with Alzheimer's disease. Those who have the courage to disclose their diagnosis to others engage in an extraordinary public service that challenges us to include the varied dimensions of people with Alzheimer's in the evolving portrait of our society.

A Lesson in Slowing Down

Betty knew that her disease could create unpredictable dilemmas for herself or for others. Having dedicated her life to understanding human behavior and interaction, she readily called upon her various skills and insights to compensate for any mistakes imposed upon her by memory loss.

I think I've learned to slow down and pay better attention. I don't want to get involved in a conversation and take a leap from here to there without being clear of that intermediate step. I want to make sure I know what I'm talking about. My caution does, in some instances, slow my response because when I am spontaneous, I have to be sure that I don't put my foot in my mouth. I might say something to person A that really was meant for person B. And that's not necessarily a good thing! You commit a few more faux pas than you do ordinarily. So sometimes you have to cover up. For example, I've learned to be a little more praising of people. I will find something to praise in someone as a means of disarming them about mistakes that I might make. It somehow makes them less upset if I lose track. I've discovered that this is very helpful. It's amazing what you can do to protect yourself!

In a way, Alzheimer's is a learning experience. I used to be able to retain a lot of information that could easily be recalled at any time, and that ability is diminishing. I'm learning that I can't always rely on that information. So instead, I may answer someone's question with another question. We all do that. It's a process of trying to help ourselves.

Also, I'm observing myself and other people a little more closely. I've always been very sensitive to body language, emotions, and attitudes. I can tell from how a person moves whether it was a good thing or bad thing that I said. I have to use my intuition a lot more than I used to in order to pick up on the meaning of what people are saying to me. These are skills that I had training in as a social worker, so at least I'm prepared.

Betty and I laughed at our mutual recognition of these communication tactics. She expressed no embarrassment in divulging her secrets. Rather, she seemed amused by her own ingenuity. She possessed an inordinate ability to solve problems, and she maintained elegant social skills. Although everyone tries to maneuver out of an occasional blunder, Betty's conscious use of her particular tactics was creative and strategic. I was refreshed by her candor because it conveyed an acknowledgment of personal pride and her resourcefulness in maintaining her self-esteem. She would not be made to feel inferior by her disease. Her impairments were very irritating and unsettling, but she tried at every turn to outsmart them. When this was not possible, she worked to be rational and accept them.

The Whole Human Being

Betty had a broad perspective on Alzheimer's disease. She felt the effects of the disease personally. As a social worker, she also placed herself in a larger social framework and was aware that she would, in time, become a recipient of the services of her own profession.

People are beginning to understand that there is such a thing as Alzheimer's disease, and the financial cost of it is becoming a big issue. There is a great deal of focus on this. There are a lot of poor people living longer and growing older, so some people are very upset about the expense of it all.

I think the most urgent issue for everyone is to learn the whole business of acceptance. I've seen too many health care professionals who have never made it to that phase. They're busy wanting to climb up to the next rung on the ladder. That's very human. I don't blame them. But they don't really accept the significance of illness for people. They know the diagnosis, but they don't take time to find out what it truly means for that person. This casualness with which professionals deal with Alzheimer's is so painful to see.

Acceptance is hard to teach. In order to learn to accept other people's deficiencies, you have to first be able to accept your own. A person with Alzheimer's disease is many more things than just their diagnosis. Each person is a whole human being. It's important to be both sympathetic and curious and to have a real interest in discovery about who that person is. You have to really be willing to be present with the person who has Alzheimer's. But there are some people who don't want to learn, and it's the looking down on and being demeaning of people with Alzheimer's that is hard to watch.

Two years later, I drove the now familiar route to Betty and Kurt's home. Betty had joined the Alzheimer's support group at the time of its inception, shortly after our first interview. She and Kurt had recently hosted the two-year anniversary party for the group members and their families. Although I had seen Betty nearly every week for the past two years, I looked forward to the chance to talk with her privately.

When I arrived, Betty and Kurt were engrossed in planning a two-month summer trip to Europe. Maps were strewn on the dining room table, and Kurt was organizing the details. Although the two had once been an equal team, the responsibility for the logistics of the trip now fell to Kurt. But this did not diminish the spirit of partnership they displayed while discussing the proposed plans for the journey. Kurt had to cancel an overnight ferry trip for lack of an available sleeping cabin. The only option was sleeping on the deck, and he had ruled that out.

"Kurt! You're getting old!" Betty bantered. "Sleep on the deck!"

Kurt looked at Betty in a pleased but incredulous way, "You're serious, aren't you?"

"Of course I'm serious!"

"You'd really do it, wouldn't you?"

Betty's audacious ambition seemed reassuring to Kurt. Her long-standing gumption had not surrendered to Alzheimer's disease. Kurt glanced at me, affirming the moment with a grin: "I think she'd really do it."

I felt warmly engaged and inspired by their conversation. Kurt and Betty, both 80 years old, were experienced rustic travelers. Last year's trip was through Alaska in their Volkswagen bus. I have a similar love of adventure, probably inherited from a few of my own spirited ancestors, and I hoped that I too would be on a ferry deck in my sleeping bag at the age of 80.

The logistics of traveling with bulky sleeping bags were difficult, and in all likelihood, the option was unrealistic. But Betty's fortitude and flexibility were invigorating. Although she was fairly realistic about her impairments, she did not want to consider her life unduly constricted by any constraints of getting old, let alone the limitations imposed by Alzheimer's disease. As we sat down in the living room to talk, Betty expressed her enthusiasm about the travel plans.

I'm excited about this upcoming trip. You won't believe what we're going to do. We talk about everything. If I said no to something, Kurt would be quick to cut if off. But I'm right with him. I hope we'll have quite a few adventures. We are still doing practically all of the things that we were doing before Alzheimer's, and I think that's important. I'm open to new things, and I'm learning all the time. It's better than shutting up and lying down.

I think my disease is a slow process, and Kurt helps me to enjoy everything there is to enjoy. We both keep in shape. We go to swim class three times a week, and during the summer we're in the ocean every day. I was thinking today when we were swimming just how nice it felt to really be physically active. I'm 80, so I can't live much longer. My life expectancy is not that great. So I don't feel cheated by the disease or like life is playing too many tricks on me. As long as one is able to maintain a relatively normal relationship in public, what does it matter if you have Alzheimer's? If I can be an ordinary human being, let me be one. Why should I worry about everything that could happen? I just deal with each problem as it comes up. I'm pragmatic. Being pragmatic is my religion.

Betty was pleased with her physical strength. In light of her cognitive changes, she took pleasure in a mastery of her physical body and celebrated her overall health. Her vital appearance and her involvement in daily activities helped her to feel inconspicuous. Although truly extraordinary, Betty strove to be ordinary—to move within her familiar circles and maintain an evolving relationship with life. The recognition that she had lived a long and rich life so far diminished the sense of injustice many younger people with Alzheimer's feel when the disease intrudes upon their careers, their plans for retirement, or their vision of a future. But appreciation did not diminish her intent to make the remaining years vital as well.

Betty's relationship with Kurt had settled down from the episodes of impatient turbulence that had rocked it a few years ago. Whether on the upcoming trip or at home, she knew she was in good hands.

We love each other. We have for 55 years, and we're not going to change. That in itself is so consistent and relaxing. I'm comfortable with Kurt because I feel he takes into account whatever deficiencies I have, and I'm learning to be dependent on him. He's also more accustomed to the memory problems now. He understands and he's helpful. When I forget things, he reminds me and he's not nasty about it. If I'm in a conversation with others and I lose something, he fills in unobtrusively. He's loving and caring. That's the important thing. This is one time that I'm glad I'm married to a social worker. I'm lucky to have him. I'm luckier than most in my position.

Kurt likes the fact that I still have a sense of humor and we can laugh about things. We can talk things out, which is terribly important. I don't hold it all in. I used to hold things back because I didn't want to put a burden on him. But now I'm more open. If something comes up, I'll talk. He also understands when I just want to remove myself sometimes and go upstairs to get away for a bit.

I think that at the point when I feel like I'm really a burden to Kurt, I'll suggest that I go into a home. It would be too much of a burden to me to be a burden to him. I don't know what the circumstances would be, but I think I'll know them when they come along. I'm not ready for a home yet, though.

Although Betty did not feel it was a significant problem, she had begun to experience bouts of despondency over her sense of being a burden to Kurt. At times, she felt she could no longer contribute meaningfully to the relationship. She continued to be deeply sensitive to the needs of others and astute to the nuances of human relationships. Yet, standing in the light of her acknowledged deficits, she tended to relegate her strengths and contributions to the shadows.

Betty had a long-standing interest in group work, and as a psychiatric social worker she had facilitated groups for people with serious problems. The Alzheimer's support group was the first she had ever joined for herself, however, and her commitment to the group was unwavering.

I'm all for support groups for people with Alzheimer's disease. That's one of the best ways to find out how varied this disease is. People are in all different stages, and if you just open your heart and your brain to it, you can learn an awful lot. If people are in the earlier stages, they can talk about their experiences and express their feelings and anticipation about the future. The primary value is in sharing experiences about a common issue without having to put up a wall because you're concerned about how people are going to respond.

The main issue is to help people to be open about Alzheimer's—not to privatize it, especially within the family.

Very often the tendency with something like this is to hold it in and suffer with it. But it isn't necessary to suffer alone. People with Alzheimer's are curious about what all of this is going to mean to their lives, and if they can get some sense of this through a support group, then they can move into this process more at ease. That's very important.

I think the nicest thing for the group is your concern as a facilitator for each individual, each special problem, and how to help. This has an effect that says, "See, if something goes wrong with you, I'm here." I think that's wonderful. When I'm in the group, I still try to help facilitate when I can; I try to talk to someone afterward, or hold a hand.

Our support group has held up remarkably well. It's nice to have the feeling that you're all in the same boat. The weekly continuity is an important part of the process. I like the way people get really concerned about someone who is missing and what's going on in the person's life. That's real grouping. It hurts me to see some people in the group who I know and care about go down faster than it seems they should. It's sad. But I'm there. I'm a part of the group, and that's it.

Whether through a group or other supportive means, our ability to count on someone caring about us is fundamental to our capacity to cope with adversity. The fear of illness is imbued with the fear of vulnerability to emotional and physical pain. Although others cannot always remedy every problem that arises, the reassurance that we will not face the challenge alone assuages our vulnerability and eases our fear with the therapeutic effects of human connection.

The support group was a place where members talked not only about Alzheimer's disease but about the other dimensions of their lives—their histories, families, travels, and interests. The more sobering milestones of progressive illness were less dominant and imposing obstacles when interspersed with brighter markers along life's path.

A Garden's Gift

While Betty and I continued our conversation, the light in the living room moved across the windows, the shadows shifted, and I became aware of the passage of time. I glanced over at the tape recorder, thinking that it should have clicked off by now. The recorder lay on the table with its light on, but on further examination, the tape had stopped winding quite some time before, leaving the bulk of our morning's conversation unrecorded. I had decided not to take notes during these lengthy interviews because note-taking can inhibit both people's ability to relax into a natural dialogue. I could have summarized our discussion, but I wanted Betty's specific words, not mine. And Betty's thousands of words were now gone. I saw no other way to remedy this significant loss but to come back some other time and start the themes of our conversation all over again. I mulled this over silently as we both sat looking at the inert instrument.

"Can I get you some coffee?" Betty's voice was undaunted. She would not be cowed by this discouraging display of defeat. I smiled at her thoughtfulness, her resiliency, and her cue that she did not want our visit to end, especially on a frustrated note.

"Water, Betty. I would love some water."

As Betty went to the kitchen for refreshments, I felt my psyche begin a dramatic shift. Throughout her life, Betty had not been easily defeated by adversity. Now, self-absorbed in my loss, I was humbled by how graciously her long-standing determination was incorporated into her daily confrontation with the potentially defeating nature of Alzheimer's disease. She knew when to step away from frustrating events—when to stop clinging to an impossibility and instead reach for a new opportunity. While I mourned over this singular lost transcription, Betty experienced such casualties on a regular basis. The details of even the briefest conversation with someone could be lost to a few moments of passing time. She often retained the gestalt but very rarely the exact verbal content. As the fog of my attachment to our verbatim conversation lifted, I awakened to the extraordinary amount of ongoing renunciation that must be exacted from one who experiences memory loss. Yet when the exact memory of conversational text is erased, can something else be revealed?

Relinquishing my loss, I transferred my gaze away from the tape recorder and looked out the living room window. Bright orange California poppies were luminous and energizing in the garden beyond. When Betty returned with water, I suggested a walk out the side door toward the vibrant color. Betty, pleased that I'd recovered from my momentary stupor, was delighted to oblige.

Betty knew the plants in their garden the same way she often knew people—not by name but by a familiarity of manner and presentation, an association with their repeated presence in the same context again and again, a reference to history imprinted in long-term memory. Although these pockets of stability dotted the garden, there was an overall randomness to the landscape. There were new plants and colorful weeds that Betty joyfully discovered on our walk, as well as old familiar specimens that she greeted with fresh and spontaneous delight. She was candid in her disapproval of some plants' tendencies to grow effusively, without restraint, but also seemed pleased by their rebel tenacity and vigor. The garden was an expressive environment, weedy and wild, blooming and reseeding itself. It was practical in that it survived with very little maintenance but could be nurtured and worked when time allowed.

As we walked, the presence of Alzheimer's disease gave way to companionship and the invigorating spirit of nature. In our mutual observation of the plants and flowers, we were engulfed in the discovery of a landscape that held us equally engaged. As a professional, I could limit my role with Betty to assessments, recommendations, and interventions. I could know her only as a patient with Alzheimer's disease whose deficits posed an array of problems to solve. Yet as Betty had so keenly remarked, she was a great deal more than her disease. She maintained a diversity of life experiences and skills that graced her personal landscape. Her identity was not defined by Alzheimer's disease any more than the garden was defined by any one plant that grew within its perimeters. She fertilized aspects of her life to bloom so brightly that one was attracted to her social graces, her adventurousness, and her candor and made less aware of her disease.

Alzheimer's was a new species in the garden, and in time, it would have a greater and greater presence. It could dominate, but it would never be the whole landscape. Some expressions of her character would lose their bloom, while other dimensions of being, like unpredicted new species, would emerge spontaneously. The pathways of our present walk would be detoured in unpredictable directions, and the seasons would mark their passage on her ever-changing garden. In years to come, the landscape could look very different, but it would still be her own.

We paused on a lower pathway to look at an old redwood tree perched on the canyon rim that overlooked the hillside and the ribbons of freeway farther beyond. I marveled at its solitary stance in arid southern California, as I was accustomed to seeing redwoods grow in lush, damp groves in my native northern California surroundings. The tree, while somewhat misplaced, exhibited an adaptability that was commendable. Betty beamed as she acknowledged it. "We had it cut back," she remarked, "and it bushed out nicely." She appreciated the tree's performance, and I imagined an affinity she felt with this sentinel on the hillside. In this random garden where surprises in life were either an impromptu offering or a new problem to be solved, she celebrated the complexity and challenge of existence. Reaching her arm out triumphantly, she saluted the presence of the redwood tree once more before resuming our walk: "It's taken a beating over the years, but it's alive! That's the important thing."

Consuelo

"Without being positive, there is no life."

We have much left to learn about Alzheimer's disease. As I looked across our conference room table at Consuelo's softly rounded and expectant face, I wished there were a different reason for our first interaction. At age 33, Consuelo, accompanied by her husband and her adoptive parents, waited to receive the results of her predictive genetic testing for a rare form of familial early-onset Alzheimer's. Our introductions were brief as we positioned ourselves around the large oblong table and awaited our neurologist's opening words.

I met Consuelo's sister long before I met Consuelo. Marta had been a patient in our University Medical Center's Huntington's disease clinic, where I served as the clinic's social worker. Her unusual jerking movements and the progressive dementia that struck her and other family members at such an early age pointed to this genetic, degenerative neurological disorder as the most likely cause of her unfortunate symptoms. Her mother developed dementia in her mid-30s and died by age 50. Her maternal uncle suffered the same condition.

But over the course of Marta's repeated visits to the clinic, her symptoms were increasingly uncharacteristic of Huntington's disease. Subsequent genetic testing for Huntington's confirmed that she did not carry the gene associated with this disease. One of our research center's neurologists, seeking to discover the condition responsible for her unusual symptoms, suggested the uncommon, but diagnostically useful, procedure of a brain biopsy. Although a brain autopsy may be done to verify the cause of death, a brain biopsy is warranted when it can help determine the cause for an unusual presentation of clinical symptoms. Confirming our neurologist's hunch, Marta's brain biopsy revealed the characteristic plaques and tangles found in the brains of people with Alzheimer's disease. Once the disease was determined, Marta underwent genetic testing to find the possible cause of her unusually early-onset disease. Diagnosed in her mid-30s, Marta joined the roughly 1 percent of people whose Alzheimer's disease is attributed to a devastating and inherited genetic mutation on chromosome 14. It now was clear that her mother had not died of Huntington's disease but of this familial form of Alzheimer's. Sadly, she passed this same fate on to her daughter.

In families with the chromosome 14 mutation, the carrier of the gene, whether father or mother, has a 50 percent chance of passing it on to the offspring. It is an alarming game of chance with very inflexible rules: if the child inherits the gene, it all but guarantees that somewhere between ages 30 and 50, symptoms of Alzheimer's disease will begin. But if the child does not inherit the gene, the risk of developing Alzheimer's disease is no greater than that of the general public. Because each birth carries the same odds, some children of a carrier will be born with the gene mutation and thus be destined to develop Alzheimer's in early to midlife, while others will live on into advanced age, free of this unfortunate inheritance.

Scientists have discovered other gene mutations on chromosomes 1 and 21 that cause the same definitive outcomes as the chromosome 14 mutation. All these forms of familial Alzheimer's disease are early-onset (occur prior to age 60) and account for far fewer than 5 percent of all Alzheimer's cases. Although the presence of these specific gene mutations can definitively predict whether one will develop early-onset Alzheimer's disease, other genes related to later-onset Alzheimer's are less absolute and only establish elevated risk. ApoE is a protein that, in the blood, helps to carry cholesterol and fat. Found on chromosome 19, one form of the gene for this protein, ApoE4, is present in up to half of those who acquire Alzheimer's after age 60. This gene, which is also influential in heart disease, may be inherited from one or both parents. Although one can have the ApoE4 gene and not acquire Alzheimer's, its presence increases the risk of developing the disease later in life.

The role of genetics in Alzheimer's disease is an intensely active area of research that holds potential for new diagnostic measures and therapeutic breakthroughs to prevent or postpone disease onset. But for those families enduring the risk of genetic disorders, life is fraught with complex challenges and decisions.

When Consuelo's family thought that their inherited illness was Huntington's disease, she underwent predictive testing to determine whether she carried the gene for that disease. The findings were negative. Although freed from the fear of acquiring Huntington's disease, Consuelo did not feel completely at ease. "When I knew there was an illness that ran in my family, I wanted to investigate it," she later recalled. "When I got the test results back on Huntington's, I wanted to believe that I was free and clear of genetic disease. But part of me didn't think I was because at the time, they didn't really know what disease my mother had."

Unfortunately, only two years later, with her sister's test confirming a completely different familial gene and disease, Consuelo once again faced the decision of whether to undergo predictive testing. If she inherited the chromosome 14 mutation, she would follow the course of her other affected family members and very likely have symptoms of Alzheimer's by the time she was 40. Because she was newly married and planning a family, the risk of having Alzheimer's disease when raising young children, combined with the 50 percent chance of passing the gene to her offspring, weighed heavily in the couple's decision to have children. After extensive counsel with a medical team who reviewed with her and selected family members the implications of genetic testing, Consuelo decided to proceed. Now, awaiting the results, there was little way to soften the news. Her test was positive: she carried the gene.

Consuelo's response was most notable for its silence. Her eyes welled with tears, and she was visibly shaken. Yet it was difficult to tell from her quiet, stunned demeanor how she was processing the information. While her husband looked devastated, her adoptive mother took the floor and asked the bulk of questions about her daughter's test results. As we addressed her questions, I wasn't certain Consuelo was hearing the answers. She was present, but her attention seemed to waver, and I wondered if she was beginning to shut down. The profound implications of testing positive for this gene were too much to take in at one meeting. We could not be the bearers of this information, then simply send her away. While our neurologist enrolled Consuelo for annual follow-up in our research center, I made arrangements to meet again with her and her husband, Juan, within the month.

Over the next year, I had six more meetings primarily with Consuelo alone. Fearing loss of privacy and a paper trail of medical records if she sought counseling through her health insurance, Consuelo talked about her concerns in the safety of our confidential research center. She was very isolated. Where did she fit into the Alzheimer's experience? She was exceptionally young and not yet diagnosed, but she lived with the knowledge that she would soon develop the disease. The prevailing image of Alzheimer's disease did not include a face like hers. Given the extraordinary impact of the disease on her life, Consuelo thought it should. She decided to be interviewed with her real name and some personal information concealed to maintain confidentiality.

There are reasons that I choose to interview people about Alzheimer's disease in their homes. When we see participants at the research center, we are in an environment that focuses on disease. We do our best to honor and address other dimensions of our participants' well-being. Nevertheless, our purpose in being a research center and their purpose for coming to see us are the same. A meeting in the home, however, opens the parameters beyond illness to include other aspects of existence; it can validate the myriad other dimensions of who a person is and what is meaningful in daily living. Thus, a patient with a disease becomes a person with a unique life story.

A few years ago, before I changed residences, I could have walked from my house to Consuelo's. I didn't know her then. But chances are fair that over the years, we passed by each other while doing errands in the neighborhood, going to the movie theaters, or shopping at the local grocery store. These are the ordinary routines of life that unite us as members of a community—the common public contexts for the private dimensions of our personal lives. We all walk shared ground, usually quite unaware of the personal aspects of one another's lives that form our inner worlds. Outwardly, when passing her on the street, one would not notice any significant changes in Consuelo over the previous year. But while external appearances in the neighborhood carried on, Consuelo's interior image had changed radically.

I parked on a familiar street in front of her home and ascended the steps to her front porch. Consuelo was expecting me, and her warm smile peeked out from her front doorway. She had just fed their pet rabbit, and as I entered the home, I was greeted by the soft rustle of lettuce. Juan accused her of feeding the rabbit too much, but it was Consuelo's natural inclination to care for things—to make sure they were without want.

The dining room table served as the focal point for the business of life in the household; shortly before I arrived, Consuelo had been sitting there writing down thoughts in preparation for our meeting. As I sat down near her, preparing to ease into our talk, Consuelo's tears preceded her voice: "I cried when I was writing this," she began. "It was bringing back old memories."

Indeed, trauma infused Consuelo's recent and remote past. As we began to recall the origins of our first meeting and the news of her test results, the links to her childhood memories were an unfortunate and unbroken chain.

A Self-Protective Response

On the day I received my test results, I heard something totally different. I didn't realize that the results were so definitive. I thought even with the gene mutation, there was still only a 50-50 chance of getting the disease. I knew that I had the chance of having Alzheimer's. But I'm more educated than the rest of my family, so I thought maybe I wouldn't get it. Then I recalled that my uncle was a lawyer and he had the disease. So, your schooling doesn't matter. You can go to college and still get Alzheimer's.

I probably clicked off when I heard the news. I was too nervous. My brain just didn't want to hear it. Many months later, when the news really sunk in, I felt like my world was falling apart—my hopes and dreams for my future. It was bad enough to have gone through my childhood. And then to find out that there was the probability of getting Alzheimer's, I felt like it just wasn't fair. But life isn't always fair.

Indeed, Consuelo's childhood had been horrendous. She was born in San Diego, the youngest of eight children. By the time she was eight years old, her mother was in a nursing home with advanced Alzheimer's disease. When Consuelo reached the sixth grade, her father was sent to jail for the gross abuses he inflicted on his children, and that was the last she saw of him. She went to live with Marta and her husband in a small town just south of the Mexican border. The family moved their mother to Mexico with them and tried to care for her in their home, but eventually they had to place her in an Arizona nursing home. Although Consuelo visited her mother occasionally, the experience was extremely painful for the young girl.

My memories of my mother before Alzheimer's disease are vague. But I can remember her being a very nice mom. I can always see in my mind a picture of her face with her curly hair. But I don't really remember all the details of those times as a child. I think that it's good I don't remember everything because the past was heartbreaking, and maybe too hard to handle. I probably coped by erasing that time from my memory.

While Consuelo lived with her sister in Mexico, she crossed the border every day to attend school in her native United States. Through the extraordinary goodwill of her junior high school teacher, Consuelo's life took a significantly brighter turn.

I had to ride a lot of buses in order to attend my junior high school in Arizona, but that's where I met my adoptive mom. She was my teacher. I was in about seventh or eighth grade then. I don't know how it came up, but I must have told her how I was getting to and from school each day. She said, "Well, maybe you can come stay with me for a while. But let me talk with my husband." I don't think I remembered the part where she said, "Let me talk with my husband," and I came back to school the next day with all of my clothes! So I moved in with them. They didn't formally adopt me until I was 18, so there was no chance that my biological father could contest the plan. My adoptive parents saved me. I would have ended up uncaring, selfish, and in trouble with the law. I think I'm a good person now because of them.

But I understand that because of the gene, I'll get Alzheimer's disease and it could start as early as my late thirties. I'm 34 now. My mom is very positive and very caring. She says, "God is with you, so don't worry about it, and if you get it, I'll take care of you." And I think, "But you already took care of me enough!"

Another Frightening Prospect

Consuelo had weathered extraordinary challenges to become a very capable and compassionate person. She felt much gratitude to her adoptive parents and had hoped to care for them in their advancing years, as they had cared for her. Having pursued a profession that enabled her to help others, she feared becoming dependent and losing the feelings of autonomy and self-worth that had been so hard to achieve.

I went to college and received my bachelor's degree in early childhood development. I wanted to take care of kids and be someone who makes a difference in their lives—a good difference. I work with developmentally disabled children in a special education program doing tutoring and life skills training. I give a lot of love to my kids because I didn't get it as a child. Until my adoptive parents, I never heard a lot of "You'll be all right; I care about you; I love you no matter what." These words are so important. All kids need them. While they are in our program, I'm like a mother to them. They call me mommy sometimes. It's a term of endearment. I get a feeling that I helped these kids through a hard time, and maybe it will make a difference in helping them to become better people. I know that the kids are our future. They need good role models, good care, and love.

You can't imagine how much I worry about work and about losing the job I love. I can't conceive of doing anything else but this work. I have to be careful who I talk to about this Alzheimer's gene, because I could lose my job. If they find out, they will be really nitpicky. I know that, eventually, they'd get rid of me. I would also lose my health insurance. My husband gets his health insurance through my employer. He doesn't make as much money as I do, so there is no way that we would be able to support ourselves on just his income. I'm sure that any disability compensation I'd receive would not be as much as my current salary. I've worked hard to be independent and to have my own health care. I don't want to feel like I'm begging for benefits.

Consuelo's emotional and financial welfare were considerably invested in her profession. As she spoke of her work, her enthusiasm and heartfelt commitment were palpable; the children engaged her intellectual abilities and her empathy and fostered a rewarding sense of purpose. They were also an invaluable escape from the preoccupation of Alzheimer's disease. There was no room for ruminations about future uncertainties when the many needs of the children were so immediate. But despite the value of staying focused on the present, Consuelo had issues before her that she could not long postpone or afford to ignore.

For many families facing Alzheimer's disease, legal and financial planning for the future becomes a significant issue. Estate planning—including durable powers of attorney for finances and health care, as well as the creation of a will or trust—is important to take into account early on in a progressive disease process, while the diagnosed family member has full capacity to participate. Because Consuelo was not yet diagnosed with Alzheimer's, there could be other areas to investigate: purchase of a more comprehensive disability policy in the event of her unemployment; supplemental health insurance that would go into effect if she lost her job and her benefits; a long-term care policy to help pay for future expenses. Once diagnosed, she would be automatically ineligible for these benefits.

Yet Consuelo had been diagnosed with a genetic mutation that would inevitably result in Alzheimer's disease. Would she be mandated to disclose this information? Should she be? As genetics becomes a larger factor in Alzheimer's and, increasingly, in other serious diseases, the ethical issues surrounding predictive testing or disclosure of test results are complex. Because a great deal of genetic testing speaks only to relative risk of acquiring disease, it is unclear how large a role findings will play in the screening of applicants or the provision of insurance policies. But given the profound ramifications of genetic tests that definitively predict future disease, requiring such information can have very serious and complex consequences.

The Decision to Be Tested

Had the outcome of Consuelo's test results indicated the absence of the Alzheimer's gene, it is unlikely she would feel ambivalent about knowing the findings. Indeed, it would have been an extraordinary relief to be able to proceed in life without the imminent threat of the disease. But the chance of receiving this happy news had been equal to the chance of receiving news of the far more sobering alternative that Consuelo now faced. I asked her what someone should consider before deciding to be tested.

It's important not to be pressured to get genetic testing. You might hate the person who forced you to do it. You might be resentful. It should be *you* who wants to do it. You have to make sure in advance that you know what you're getting into. You have to consider all the pros and cons of knowing the results, and see if you think you can handle them before being tested. Choose the decision that you want, not what someone else wants. And even if you're tested, you don't have to get the results. You can do it just to benefit science. If I did it all over again, I wouldn't want to know the results. It's just too hard. Nothing good has come out of it.

I decided to be tested because we wanted to have kids. If I didn't want kids, I wouldn't have been tested. I felt some pressure from Juan about having it done because of this issue. If I was positive, he knew he didn't want kids. But I was never completely sure that positive test results would mean I shouldn't have children.

I still want kids. It's always been my desire to have children. I feel incomplete without them. My family feels incomplete. It's hard when I watch TV and they have commercials of babies. I start to cry. Juan doesn't want to have kids because of the chances of passing on the gene. It's hurting our relationship. But I have to think positively that there will be something to either cure the disease completely, or at least slow it down. If I was too sick with Alzheimer's to take care of them, we have a lot of dependable family and friends who would help us.

The decision to undergo predictive genetic testing is very complex and warrants thorough consideration with a physician or genetics counselor. Unlike other genes that predict only the relative risk of acquiring a specific disease, testing positive for the chromosome 14 mutation is definitive and predicts the onset of a profound and progressive illness in one's midlife. Many of us have sometimes wondered what life would be like if we could see into a crystal ball of our future. Would we live differently knowing what might lie ahead? How would this knowledge affect our ability to enjoy the present and set goals for the future? Do we want this kind of information, or are we being pressured by others to acquire it?

Family planning was at the root of Consuelo's decision to be tested. Juan thought they were in agreement that they would not have children if she tested positive for the gene. But once the test result was known, any previous understanding the couple had about its influence on family planning was unraveled by Consuelo's grief about the prospect of remaining childless.

Consuelo realized that once her symptoms began, the progression of Alzheimer's would, in time, render her incapable of caring for children. She did not want to subject her children to the same trauma she experienced of losing a mother to Alzheimer's disease. Yet even as we spoke of her responsible and realistic concerns, she provided a compelling rebuttal that argued the value of her life and her capacity to survive obstacles. In spite of a traumatic childhood and the imminence (barring prevention or a cure) of Alzheimer's disease, Consuelo had overcome enormous obstacles to become a very caring person who contributed a great deal to the lives of others. By choosing not to have children for risk of what they might endure, was she saying that her own life had not been worth living? Consuelo's positive thinking convinced her that even if she did pass on the gene, there could well be a cure for Alzheimer's by the time the child was old enough to express the disease. So why deny life now for fear of the future?

Some may not share Consuelo's perspective, however. For a spouse, the economic and emotional hardships of raising a child as a single parent or even with the help of other family members can seem incomprehensible when facing the prospect of caring for a disabled partner. Some question their own ability to cope with their loved one's illness, let alone how a child would make it through. Others think it cruel to bring children into the world knowing the 50-50 odds of passing on the gene, and doubt a cure will be forthcoming in time for the next generation.

Although childbearing is a personally, socially, and ethically complex issue, it is not a concern for the vast majority of *families facing Alzheimer's disease. Yet there is something more universal in the Alzheimer's experience at the core of Consuelo and Juan's turmoil. The demands of Alzheimer's are fraught with paradoxes—for the person with the disease, for the caregiver, and for those who provide services to the family. We are asked to be both rational and compassionate, pragmatic yet intuitive; we remain steadfast while trying to be flexible, realistic while ever hopeful. In the best of circumstances, we find a way to balance, validate, and use each component. Yet when faced with a primary problem or decision, we are likely to lean more toward one side or the other of these seemingly contradictory stances and define a style of coping that is influenced by our approach to life. Sometimes people's coping styles are complementary; at other times, they clash.*

Survival Skills

Consuelo did not deny the gravity of her circumstances; she wanted reprieve from them and tried to stay focused on hopeful possibilities. Because she was unable to affect the pending onset of disease, she sought other avenues in life where she could have choices and make decisions.

I'm a positive thinker. If I wasn't, I'd probably be heaven knows where, especially given my childhood. I had to learn to be strong. I think my upbringing helped me to be able to cope with bad things, and I consider Alzheimer's disease pretty bad. I think about situations that occur in my daily life and I ask myself, "Can I fix this, or not?" I look at what I *can* do and that helps me decide how to proceed. It's a survival skill.

I think a lot about getting Alzheimer's, and I do cry about the thought. But most of the time I try to be positive. If I see myself crying, I think, "Don't do that!" Because without being positive, there is no life. We have to live for each day and enjoy it as best as possible. I feel grateful that I'm alive, that I have family and friends. We have a place to live and secure jobs, so we don't have to worry about living out on the street. People who take these things for granted didn't go through what I did as a kid and aren't living with what I am experiencing now.

My beliefs help me, but I have to keep reminding myself of them. *(Consuelo smiles at her confession.)* As anyone would, I've wondered, "Why me? I'm a good person, so why is this happening to me?" But only God knows. I'm not angry, but I feel helpless because God can't answer my question with a human voice. But I realize he knows what he's doing and he'll take care of me. He'll only give me as much as I can handle. We all have our trials; everybody goes through something that is difficult. I pray often for the strength, and for my husband to have the strength, to go through this. I have to remember that my husband and I are only human.

I'm sleeping better now, but Juan isn't sleeping too well. He'll be up sometimes at one or two in the morning watching TV because he can't sleep. He seems depressed and then other times not. But I am probably the same way. I'm on Prozac. It helps a little, but it doesn't help as much as I thought that it would.

I try to talk with him, but he doesn't talk. It's so hard to get things out of him. He just keeps it all inside. When I verbalize what I'm feeling, it lets out the stress and I feel less bottled up. Then at least for a while, I feel better.

Juan pays a lot of attention to me. He's kind and has a good heart. I know that I can count on him for now. But I don't know if he'll be able to handle the situation in the future. I worry that he'll resent me. He'll probably get tired of taking care of me, and I wouldn't want to be a burden to him. I wouldn't want him to have to go through that. But we enjoy our time together now. We like to go camping; we go to the movies; we enjoy taking walks and going out with friends. When Alzheimer's disease comes into my mind, I just try not to think about it. I say, "God, take care of it, I can't handle it."

None of us can truly predict our futures. Although some things may seem certain, life involves countless unpredictable variables well beyond our control. But because Consuelo knew a profoundly influential component of her destiny, it was as if she sometimes lived in two time zones. The knowledge that Alzheimer's disease was in her near future made her keenly appreciative of the present—of her relationships, her work, her faith, and the security derived from basic comforts that rarely inspire our conscious gratitude. Yet images of the future were often superimposed on her daily life, creating a disorienting and depressing perspective. Sometimes it was painfully hard to focus: even as she tried to keep her attention on the present, the prospect of Alzheimer's was an inseparable and distracting part of the experience.

The thought of pending disease seemed to foreshorten Consuelo's images of the future. Some people in their thirties are thinking ahead to hopes and goals in 5- or 10-year increments; when I asked Consuelo about any planning for the future, she discussed a short vacation she was hoping to take with her husband next month. That was as far as she dared to project her thoughts.

Looking Ahead

Often, I have a lot of things on my mind and Juan says something to me, but I'm thinking about something else and only half hearing what he's saying. So later on, he'll tell me again and I'll say, "I didn't know you said that!" But I'm not worried about my memory right now. My health is OK; I just feel depressed. Somehow or another we'll be able to handle it. I know I don't have Alzheimer's now, though, or else everyone would be panicking. But it's like a clock ticking off, and I think, "Stop it!"

I don't want to be told if there are findings in my yearly evaluations at the research center that indicate the onset of Alzheimer's. I'll let you know when I want more information. As soon as I see signs, I'll want help. If I forget where I am or where I'm going, or if I don't recognize my family or someone I should know, or if I'm at work and don't know what I'm doing, then I'll be worried. Then I'll have to accept that I have the disease, and I'll want to take any treatment that comes along.

I hate the idea that if I do get Alzheimer's, I'm going to end up being helpless. I'll have to be taken care of. It brings up sad memories about my mom. I'm just too independent, and I can't imagine having to depend on others. But you have to find people to help you through the hard times. I would try a support group, but if it wasn't helpful, I wouldn't continue to go. I'm a trier. I would want to learn ways to cope with the disease—strategies for living day to day. I don't know if it's harder at my age than it would be if I were 70. Maybe it would be the same. But I always expected to get ill or have Alzheimer's when I got older, not now when I'm in my thirties. It does make me want to do things now, like traveling, that I might have postponed until later years. I want to see London and Paris.

I have a lot of friends and family who will help me through this. People in my situation need a lot of companionship and friends. My family gives me peace: Juan, my parents, my sister Sara, and a couple of girlfriends. Just one of my girlfriends knows. I work with her and I asked her to keep an eye on me. I let her know what I was going through and asked her to pray for me and to keep it between us. She said she'd be there for me and not to worry. I have confidence in her. I trust her completely.

I hope that there is a rainbow at the end of the tunnel. I'm a survivor, and I have to trust God. I feel that there will be a cure or at least something that will slow down the disease. But please, hurry up and find it!

The urgency in Consuelo's voice resonated with me as I left her home and merged back into the bustling streets of the neighborhood. No amount of urban noise could drown out the compelling messages she delivered. Nor did I want it to. Sometimes I sought a few moments of quiet after our meetings for contemplation; at other times, I would busy myself with a mundane task, with the soothing repetition of something controllable while other thoughts simmered slowly on one of the back burners of my mind.

I dislike Alzheimer's disease. I would like to see it eradicated. I would like to meet all these people under other circumstances; I am sometimes surprised by Consuelo's willingness to seek out and continue contact with those of us who have been introduced into her life by the prospect of disease. I imagine that in trying to stay focused on the present, one would want to avoid those who represent a frightening dimension of the future.

Yet this is an extraordinary paradox in human associations: a relationship founded on something as discouraging as Alzheimer's disease is not necessarily a discouraging experience. The prospect or experience of Alzheimer's is a profound hardship for those with the disease and for those providing care; for many people, it is difficult to find much encouraging about the process. But the alienating domain of illness is often assuaged by the refuge of caring and respectful connections. Though not a cure for disease, they humble adversity to become healing triumphs of another sort.

Sometimes it is difficult to know what constitutes care and respect. How can we acknowledge such painful, frighten-

ing truths and not shatter all confidence and hope? Consuelo has said that beyond the knowledge of her positive gene status, she currently wants no more information about the absence or presence of early signs of disease. It is out of respect for her decision-making autonomy and care for her overwhelmed feelings that we honor her request. It is possible, however, that during her yearly evaluations at our research center, we will recognize subtle changes and detect the onset of symptoms long before Consuelo solicits further feedback or meets her own criteria for concern. The medicines that are currently available, while possibly offering some benefit in the early stages, cannot significantly alter the course of Alzheimer's. But there may be more promising treatments in the future. If, at that time, she has developed symptoms that could respond to medicine, care and respect may mean reevaluating her request for no further feedback about her condition so that she can receive available treatment.

Although not yet diagnosed with Alzheimer's, Consuelo is profoundly affected by the impact of impending disease on her life and is already experiencing many of the concerns families face as Alzheimer's progresses: the need to plan for the future while staying focused on the present; the family stressors when coping styles vary; the need to identify and establish supportive relationships; the attempts to balance hope and realism to maintain one's equilibrium. Unlike the other persons profiled in this book, for the purposes of this interview, Consuelo's identity is masked. We may not know exactly who she is, but neither can we know the faces of many others who will be affected by this disease in the future. In the days, months, and years to come, there will be many unidentified people, both familiar and unfamiliar, who will join the millions already diagnosed with this disease.

Although Consuelo's genetic circumstances are unusual, her voice is crucial to our collective testimonies about living with Alzheimer's. The relative rarity of her early-onset familial disease only exacerbates her loneliness and intensifies the need for community. In the years to come, through the effects of family history or the workings of chance, there will be thousands more voices at risk of isolation, each speaking both the common and unique dimensions of their minds. As this happens, we must assure them all that there will be a community of people who will continue to listen, respond, and provide reassurance that in this disease, each voice counts.

Part Three

RESPONDING

Just as the seven individuals profiled in this book have spoken their minds so that we may better understand the personal experience of Alzheimer's, so too do thousands of people around the globe work diligently to expand their minds, ask questions, and seek solutions to remedy the effects of this disease. In recent years, there has been a steady flow of scientific discoveries that have contributed to the body of knowledge about Alzheimer's. Scientists now know a great deal more about the makeup of the neuritic plaques and neurofibrillary tangles that define Alzheimer's, and discoveries in genetics have unveiled a whole new dimension of the disease. It is now thought that Alzheimer's is not necessarily caused by a single factor. Rather, a combination of genetic and physiological influences may interact differently in each person with Alzheimer's to result in the onset of disease.

As far as we know, the disease is worldwide, sparing no country or group of people. There are known factors, however, that may increase one's risk for Alzheimer's. Age, ApoE4 status, and family history or genetic predisposition are the most significant and consistent risk factors. Although once a popular notion, there is no evidence that Alzheimer's disease is viral, and it certainly is not contagious. Theories implicating exposure to aluminum have fallen out of favor, and currently there are thought to be no definite environmental or dietary influences. Many people with no recognized risk factors develop Alzheimer's, while others with one or more never develop the disease.

Prevention

Another area of significant research and great public interest is the postponement or prevention of disease onset. At present, there is some evidence to indicate that mentally rigorous occupations, continued learning, and challenging creative endeavors may delay the onset of Alzheimer's. The old adage "Use it or lose it" is too simplistic, as many very creative and intellectually active people are afflicted with the disease. But scientists are investigating whether continuing to keep one's mind mentally stimulated throughout life can help to maintain a larger reserve of healthy nerve cells to curb the destructive effects of disease for a longer period of time.

Researchers continue to debate whether estrogen is protective against Alzheimer's through its beneficial effect on nerve cell health and growth. Many studies indicate that post-menopausal women on estrogen replacement therapy have a lower prevalence of Alzheimer's disease. Other research findings, however, are not as conclusive; vitamin E and other antioxidant supplements may help to maintain cell health and ward off disease, but this premise also needs further investigation.

While scientists work toward a cure, many researchers place more immediate hope in postponing the age of onset in the aging population and in treatments that will slow the rate of deterioration in those people already experiencing symptoms.

Treatments

It was once thought that there was little reason to make an early diagnosis of Alzheimer's disease because there was little treatment to offer the person afflicted. But now scientists are investigating ways to make an accurate diagnosis as early as possible in the course of the disease. Because many new treatments aim to slow disease progression, early diagnosis allows these medicines to be administered as soon as possible, when their effects may be the most potent and beneficial. If symptoms can be detected and arrested early, the body may be able to combat the disease effectively for a longer period of time.

Over the past decade, treatments for Alzheimer's have largely focused on drug therapies that can maintain essential levels of the neurotransmitter acetylcholine. Neurotransmitters are the chemicals in the brain that relay messages between nerve cells. Acetylcholine is essential for memory functioning and is significantly reduced in the brains of people with Alzheimer's disease. In recent years, drugs such as Cognex, Aricept, and a number of other similar compounds have been shown to be effective in reducing the breakdown of this neurotransmitter in some people. But not everyone responds to these drugs, and those who do experience only a small positive effect for a limited period of time.

Other treatments under investigation include antioxidants, anti-inflammatories, and nerve growth factor. Antioxidants (such as vitamin E) inhibit the formation of unstable molecules (free radicals) that can contribute to nerve cell damage; anti-inflammatory drugs may be beneficial in reducing the inflammation that can occur in the brains of people with Alzheimer's disease; nerve growth factor (a substance produced in the brain to help cells work effectively) may have potential to prevent cell death and arrest disease progression.

There are many dimensions to the complex impact of Alzheimer's disease on the brain. Therefore, some scientists feel it is unlikely that any one medication will be the magic remedy. Rather, treatment of Alzheimer's may include a combination of therapies that maintain levels of critical neurotransmitters, reduce brain inflammation, enhance overall nerve cell health, and limit cell death. New clinical drug trials designed to meet these goals are under way throughout the United States and

155

around the world. Scientists can understand the potential benefits of these therapies and develop new, more effective compounds only when individuals with Alzheimer's volunteer for these studies. This teamwork between investigating scientists and Alzheimer's patients is essential to advances in research to find solutions to this complex disease.

A Collective Effort

As we rely on those with Alzheimer's disease to participate in our hopes and discoveries, so too do they rely on us to participate in theirs. Recently, in the support group we facilitate for people with Alzheimer's, group members discussed the issue of hope and what constitutes or inspires the feeling. Many referred to the important inspiration of ongoing research findings, the sustaining love of family and friends, the importance of maintaining meaningful activity, and the fact of simply waking up each day. These and many other essential experiences inspire the feeling that despite considerable challenge, there is opportunity for encouraging possibilities. One group participant's comment went further to speak of hope's fundamental role in her basic sense of identity: "Instead of believing 'I think, therefore I am', I say, 'I *hope*, therefore I am.'"

Beyond clear and cogent thought, the discovery and experience of hopeful possibilities can become the core of existence and form the fuel for being when thoughts become fragmented and increasingly unfamiliar. When memory and intellect diminish, people with Alzheimer's disease can still experience the hope that they will be treated with kindness and dignity; the hope for pleasurable experiences and relief from physical and emotional pain; the hope for connection when they feel separated by disease; the hope that we listen when they try to communicate.

The individuals profiled in this book have given us the opportunity to hear their reflections, experiences, inspirations, and concerns. If we have gained insights in this process, then we have learned the value of listening. And this is essential. For whether we inquire as scientists, families, professionals, or members of the public, the minds of people with Alzheimer's disease have many more messages that are waiting to be spoken.

Appendix:
Selected Resources

The following resources can be helpful in acquiring further information about all aspects of Alzheimer's disease.

Alzheimer's Disease Education and Referral (ADEAR)
P.O. Box 8250
Silver Spring, MD 20907-8250
Phone: 800-438-4380
Fax: 301-495-3334
Email: adear@alzheimers.org
Web site: http://www.alzheimers.org/adear

Established by the National Institute on Aging (NIA), ADEAR is a clearinghouse authorized to collect, catalogue, and distribute information about Alzheimer's disease. ADEAR responds to requests for information on any aspect of the disease, including research updates, reading recommendations, and resources for coping. They can also direct you to the nearest NIA-funded Alzheimer's Disease Research Center.

THE NATIONAL ALZHEIMER'S ASSOCIATION
919 North Michigan Avenue
Suite 1000
Chicago, IL 60611-1676
Phone: 800-272-3900
TDD access: 312-335-8882
Fax: 312-335-1110
Email: info@alz.org
Web site: http://www.alz.org

Founded in 1980, the Alzheimer's Association is a national
voluntary health organization dedicated to research into Alzheimer's disease as well as education and support for people
with the disease, their families, and caregivers. There are more
than 200 chapters nationwide. The national office can connect
you with a local chapter in your region.

THE NATIONAL ALZHEIMER'S ASSOCIATION PUBLIC POLICY OFFICE
1319 F Street, NW
Suite 710
Washington, DC 20004
Phone: 202-393-7737
Fax: 202-393-2109
Email: pp@alz.org
Web site: http://www.alz.org

This office tracks, initiates, and promotes important legisla-
tion and public policy related to Alzheimer's disease.

Alzheimer's Disease International
45-46 Lower Marsh
London SE1-7RG
United Kingdom
Phone: (44171) 620-3011
Fax: (44171) 401-7351
Email: adi@alzdisint.demon.co.uk
Web site: http://www.alz.co.uk

Alzheimer's Disease International is an umbrella organization of 42 worldwide Alzheimer's associations that offer support and advice to people with Alzheimer's disease and their families. The organization networks with international chapters, encourages research, supports an annual international conference, and disseminates information.

Area Agency on Aging
9335 Hazard Way Suite 100
San Diego, CA 92123
Phone: 800-510-2020
Fax: 619-495-5080

Funded through the federal Older Americans Act, the Area Agency on Aging has community-based agencies nationwide that provide information and referral, community programming, and general assistance to the elderly and the frail disabled. The national office can connect you with a local chapter in your region.